NATIONAL ACADEMIES *Sciences Engineering Medicine*

NATIONAL ACADEMIES PRESS
Washington, DC

Practices and Standards for Plugging Orphaned and Abandoned Hydrocarbon Wells

Linda Casola, Noel Walters, and Cameron Oskvig, *Rapporteurs*

Board on Earth Sciences and Resources

Division on Earth and Life Studies

Board on Infrastructure and the Constructed Environment

Division on Engineering and Physical Sciences

Proceedings of a Workshop

NATIONAL ACADEMIES PRESS 500 Fifth Street, NW Washington, DC 20001

This activity was supported by a contract between the National Academy of Sciences and the U.S. Department of the Interior. Any opinions, findings, conclusions, or recommendations expressed in this publication do not necessarily reflect the views of any organization or agency that provided support for the project.

International Standard Book Number-13: 978-0-309-72914-7
International Standard Book Number-10: 0-309-72914-9
Digital Object Identifier: https://doi.org/10.17226/28035

This publication is available from the National Academies Press, 500 Fifth Street, NW, Keck 360, Washington, DC 20001; (800) 624-6242 or (202) 334-3313; http://www.nap.edu.

Copyright 2025 by the National Academy of Sciences. National Academies of Sciences, Engineering, and Medicine and National Academies Press and the graphical logos for each are all trademarks of the National Academy of Sciences. All rights reserved.

Printed in the United States of America.

Suggested citation: National Academies of Sciences, Engineering, and Medicine. 2025. *Practices and Standards for Plugging Orphaned and Abandoned Hydrocarbon Wells: Proceedings of a Workshop*. Washington, DC: The National Academies Press. https://doi.org/10.17226/28035.

The **National Academy of Sciences** was established in 1863 by an Act of Congress, signed by President Lincoln, as a private, nongovernmental institution to advise the nation on issues related to science and technology. Members are elected by their peers for outstanding contributions to research. Dr. Marcia McNutt is president.

The **National Academy of Engineering** was established in 1964 under the charter of the National Academy of Sciences to bring the practices of engineering to advising the nation. Members are elected by their peers for extraordinary contributions to engineering. Dr. John L. Anderson is president.

The **National Academy of Medicine** (formerly the Institute of Medicine) was established in 1970 under the charter of the National Academy of Sciences to advise the nation on medical and health issues. Members are elected by their peers for distinguished contributions to medicine and health. Dr. Victor J. Dzau is president.

The three Academies work together as the **National Academies of Sciences, Engineering, and Medicine** to provide independent, objective analysis and advice to the nation and conduct other activities to solve complex problems and inform public policy decisions. The National Academies also encourage education and research, recognize outstanding contributions to knowledge, and increase public understanding in matters of science, engineering, and medicine.

Learn more about the National Academies of Sciences, Engineering, and Medicine at **www.nationalacademies.org**.

Consensus Study Reports published by the National Academies of Sciences, Engineering, and Medicine document the evidence-based consensus on the study's statement of task by an authoring committee of experts. Reports typically include findings, conclusions, and recommendations based on information gathered by the committee and the committee's deliberations. Each report has been subjected to a rigorous and independent peer-review process and it represents the position of the National Academies on the statement of task.

Proceedings published by the National Academies of Sciences, Engineering, and Medicine chronicle the presentations and discussions at a workshop, symposium, or other event convened by the National Academies. The statements and opinions contained in proceedings are those of the participants and are not endorsed by other participants, the planning committee, or the National Academies.

Rapid Expert Consultations published by the National Academies of Sciences, Engineering, and Medicine are authored by subject-matter experts on narrowly focused topics that can be supported by a body of evidence. The discussions contained in rapid expert consultations are considered those of the authors and do not contain policy recommendations. Rapid expert consultations are reviewed by the institution before release.

For information about other products and activities of the National Academies, please visit www.nationalacademies.org/about/whatwedo.

**PLANNING COMMITTEE FOR A WORKSHOP ON
PRACTICES AND STANDARDS FOR PLUGGING ORPHANED
AND ABANDONED HYDROCARBON WELLS**

MARY HART FEELEY (*Chair*), ExxonMobil Exploration Company (*retired*)
MARY KANG, McGill University
DONALD NATHAN MEEHAN, Texas A&M University
MILEVA RADONJIC, Oklahoma State University
JAMES ALLEN SLUTZ, National Petroleum Council

Staff

DEBORAH GLICKSON, Director, Board on Earth Sciences and Resources
CAMERON OSKVIG, Director, Board on Infrastructure and the Constructed Environment
EVAN ELWELL, Associate Program Officer, Air Force Studies Board
MILES LANSING, Senior Program Assistant, Board on Earth Sciences and Resources
NOEL WALTERS, Associate Program Officer, Board on Earth Sciences and Resources

BOARD ON EARTH SCIENCES AND RESOURCES

ISABEL P. MONTAÑEZ (*Chair*), University of California, Davis
THORSTEN W. BECKER, University of Texas at Austin
MICHELE L. COOKE, University of Massachusetts Amherst
MARY H. FEELEY, ExxonMobil Exploration Company (*retired*)
KATHARINE W. HUNTINGTON, University of Washington
KRISTEN KURLAND, Carnegie Mellon University
MICHAEL MANGA, University of California, Berkeley
W. ALLEN MARR, Jr., Geocomp Corporation
PATRICIA F. McDOWELL, University of Oregon
JESSICA P. MOORE, West Virginia Geological and Economic Survey
ANN S. OJEDA, Auburn University
DAVID B. SPEARS, Virginia Department of Mines, Minerals and Energy (*retired*)
DAVID W. SZYMANSKI, Bentley University
JOLANTE W. VAN WIJK, Los Alamos National Laboratory

Staff

DEBORAH GLICKSON, Director
EMILY BERMUDEZ, Senior Program Assistant
CHARLES BURGIS, Associate Program Officer
MAYA FREY, Senior Program Assistant
CYNTHIA GETNER, Senior Financial Business Partner
SAMUEL KRAFT, Senior Program Assistant
MILES LANSING, Senior Program Assistant
SAMMANTHA L. MAGSINO, Senior Program Officer
MARGO REGIER, Program Officer
BRYAN RUFF, Senior Program Assistant
JONATHAN TUCKER, Program Officer
NOEL WALTERS, Associate Program Officer

BOARD ON INFRASTRUCTURE AND THE CONSTRUCTED ENVIRONMENT

JESUS M. DE LA GARZA (*Chair*), Clemson University
BURCU AKINCI, Carnegie Mellon University
STEPHEN AYERS, The Ayers Group, LLC
BURCIN BECERIK-GERBER, University of Southern California
LEAH BROOKS, George Washington University
MIKHAIL V. CHESTER, Arizona State University
JAMES "JACK" DEMPSEY, Asset Management Partnership, LLC
LEONARDO DUEÑAS-OSORIO, Rice University
DEVIN K. HARRIS, University of Virginia
DAVID J. HAUN, Haun Consulting, Inc.
CHRISTOPHER J. MOSSEY, Fermi National Accelerator Laboratory
ANDREW PERSILY, National Institute of Standards and Technology
ROBERT "BOB" RAINES, Atkins Nuclear Secured
JAMES RISPOLI, North Carolina State University
DOROTHY ROBYN, Boston University
SHOSHANNA D. SAXE, University of Toronto

Staff

CAMERON OSKVIG, Director
JIM MYSKA, Senior Program Officer
JOSEPH PALMER, SR., Program Assistant
BRITTANY SEGUNDO, Program Officer
DONAVAN THOMAS, Finance Business Partner

Reviewers

This Proceedings of a Workshop was reviewed in draft form by individuals chosen for their diverse perspectives and technical expertise. The purpose of this independent review is to provide candid and critical comments that will assist the National Academies of Sciences, Engineering, and Medicine in making each published proceedings as sound as possible and to ensure that it meets the institutional standards for quality, objectivity, evidence, and responsiveness to the charge. The review comments and draft manuscript remain confidential to protect the integrity of the process.

We thank the following individuals for their review of this proceedings:

DWAYNE PURVIS, Purvis Energy Advisors
DAVID SPEARS, Virginia Department of Mines, Minerals and Energy (*retired*)

Although the reviewers listed above provided many constructive comments and suggestions, they were not asked to endorse the content of the proceedings nor did they see the final draft before its release. The review of this proceedings was overseen by Christine Ehlig-Economides, University of Houston. She is responsible for making certain that an independent examination of this proceedings was carried out in accordance with standards of the National Academies and that all review comments were carefully considered. Responsibility for the final content rests entirely with the rapporteur and the National Academies.

We also thank staff member Jeffrey Kast for reading and providing helpful comments on this manuscript.

Contents

1 **INTRODUCTION** 1
 Workshop Overview, 1
 Sponsor Remarks, 3
 Organization of this Proceedings, 4

2 **ORPHANED AND ABANDONED WELL-PLUGGING: COSTS, CHALLENGES, AND BENEFITS** 6
 Introduction, 6
 Historic and Current Well-Plugging Efforts Across the States, 6
 Orphaned Wells: Locations, Attributes, and Costs, 7
 Research and Development Program for Undocumented Orphaned Wells, 10
 Commentary from State Oil and Gas Leaders, 11
 Open Discussion, 13
 Compilation of State Standards and Procedures for Plugging and Abandoning Wells, 15

3 **EXAMPLES OF WELL-PLUGGING PRIORITIZATION CONSIDERATIONS: EVALUATING WELLBORE INTEGRITY AND SUBSURFACE CONDITIONS** 19
 Introduction, 19
 Examples of Prioritization Considerations for Managing Permanent End-of-Life Solutions for Marginal, Idled, Orphaned, and Other Wells, 19
 Orphaned Well-Plugging Prioritization in Wyoming, 20
 Orphaned and Abandoned Well-Plugging in Pennsylvania, 22

Railroad Commission of Texas: Standard Practices for Plugging
  Orphaned and Abandoned Wells, 23
Evaluating Wellbore Integrity and Subsurface Conditions in Alaska's
  Orphaned Wells, 25
Perspectives From the Bureau of Land Management, 26
Open Discussion, 29

**4 EXAMPLES OF PROCEDURES AND BEST PRACTICES
    FOR WELLBORES**   30
Introduction, 30
Challenges to Plugging Orphaned Wells, 30
Example Learnings from Gulf of Mexico Plugging and Abandonment
  Experience, 31
The Process: Evaluation, Planning, and Execution, 33
Open Discussion, 34

5 ENVIRONMENTAL RISKS AND MONITORING   36
Introduction, 36
The Environmental Protection Agency and the Quantification of
  Methane from Wells, 38
Potential Methods for Quantifying Methane Emissions from Abandoned
  Oil and Gas Wells, 41
Methane Emissions and Water Quality Impacts Around Orphaned
  and Abandoned Hydrocarbon Wells, 42
Leveraging Publicly Available Data to Understand Well-Integrity Risks, 45
Informing Groundwater-Quality Monitoring with Models, 47
Open Discussion, 49

6 REMEDIATION, RECLAMATION, AND RESTORATION   52
Introduction, 52
National Park Service Orphaned Wells Projects, 53
Surface Reclamation and Restoration, 54
Colorado's Orphaned Well Program, 56
A Modern Approach to Reclamation, 57
Open Discussion, 60

**7 ADVANCES IN PLUGGING AND ABANDONMENT FOR
    IDLED WELLS**   62
Introduction, 62
Lessons Learned from the American Association of Petroleum Geologists:
  2020–Present, 62
New and Upcoming Plugging and Abandonment Technology, 63
Urban Abandonment: Challenges and Solutions, 65

U.S. Geological Survey Science to Support Orphaned Well-Plugging: Historical Drilling, Produced Waters Geochemistry, and Groundwater Quality, 67
Isolating Annuli Using Shale/Salt as a Barrier, 68
Open Discussion, 69

8 EXAMPLES OF KEY WORKSHOP THEMES   72

REFERENCES   74

APPENDIX A: WORKSHOP STATEMENT OF TASK   79

APPENDIX B: WORKSHOP AGENDA   80

APPENDIX C: WORKSHOP PLANNING COMMITTEE MEMBER BIOGRAPHIES   84

1

Introduction

WORKSHOP OVERVIEW

When oil and gas production began in the 19th century in North America, standards and regulations for the drilling and plugging of wells had not yet been developed. Over time, many of these and other wells were abandoned—unplugged, or not plugged to modern standards, and have sat idle for an extended, possibly unknown, period of time. These wells might not have been originally operated and maintained in accordance with existing statutes and regulations and, due to degradation over time and potential improper prior operations, they can emit methane, contaminate groundwater, and impact ecosystems, creating risks for both the environment and the public. Orphaned wells are documented, unplugged, and nonproducing with no known owner or operator capable of properly closing the well. A documented well has a record establishing the existence and location. Approximately 150,000 orphaned wells have been identified on state, private, and federal lands. Undocumented wells are abandoned wells for which there is limited or no knowledge, including location, and are not in any regulatory agency's inventory or for which the regulatory agency has some evidence, but requires additional research or investigation to verify. Undocumented wells represent a vast uncertainty not just in remediating individual wells, but also for estimating the cost of addressing the problem nationwide. This uncertainty is highlighted by the wide range in the number of estimated undocumented wells specified in the presentations from 250,000 to 800,000. For the purposes of this workshop and proceedings, "wells" refer to hydrocarbon wells, unless otherwise explicitly specified.

As new streams of funding continue to emerge, states are expanding their efforts to address the health and safety concerns presented by these wells, while encountering unique challenges and opportunities related to the wells themselves, surrounding areas, and broader well-plugging requirements. To explore and share the variety of existing procedures and standards for plugging orphaned and abandoned wells, including current best practices for well-plugging technologies, the National Academies of Sciences, Engineering, and Medicine convened a workshop on July 18–19, 2024.[1] Sponsored by the Department of the Interior's (DOI's) Orphaned Wells Program Office (OWPO), the workshop included members of the federal government, state leaders, tribal representatives, industry experts, and other affected parties. This proceedings has been prepared by the workshop rapporteurs as a factual summary of what occurred at the workshop. The planning committee's role was limited to planning and convening the workshop (see Appendix C for biographical sketches of the workshop planning committee members). The views contained in the proceedings are those of individual workshop participants and do not necessarily represent the views of all workshop participants, the planning committee, or the National Academies of Sciences, Engineering, and Medicine.

After welcoming the attendees to the event, workshop planning committee chair Mary Feeley, formerly of ExxonMobil, explained the goals of the workshop:

- Discuss historic and current well-plugging standards as well as design and operational practices used in the United States;
- Assess how standards and practices differ based on well age, well depth, well location, material specification, geologic and geophysical environment, production type, distance to populated areas, and remediation and restoration requirements;
- Consider how cost, technology, and other factors impact well-plugging plans; and
- Examine the environmental benefits of and the mitigation of adverse environmental impacts from well-plugging (see Appendix A for the workshop's full statement of task and Appendix B for the workshop agenda).

Feeley indicated that in addition to supporting the work of both OWPO and the individual states, the information gathered during this workshop will help inform an upcoming National Academies consensus study, Technologies and Practices for Plugging and Remediating Orphaned and Abandoned Oil and Gas Wells.[2] The consensus study will examine the following:

- Current and emerging plugging and abandonment technologies, best practices, equipment, and materials for well characterization, wellbore plugging and barrier placement, wellbore integrity and verification, and durability and lifespan;
- Evaluate unexpected or unique circumstances that necessitate varying criteria and standards, including engineering design, costs, logistics, and technical management;

---

[1] To view videos of the workshop presentations and discussions, see https://www.nationalacademies.org/event/42861_07-2024_practices-and-standards-for-plugging-orphaned-and-abandoned-hydrocarbon-wells-a-workshop.

[2] For more information about the consensus study, see https://www.nationalacademies.org/our-work/technologies-and-practices-for-plugging-and-remediating-orphaned-and-abandoned-oil-and-gas-wells.

- Assess available data on potential causes, frequency, consequences, and remediation of plug failures;
- Examine post-plugging monitoring techniques, approaches, and technologies that are or will be important for the long-term protection of both the environment and public health and safety; and
- Identify technology, materials, and policies that warrant further research and that could contribute to the success of well-plugging and abandonment efforts by industry, states, tribes, and federal agencies.

The consensus study report could inform a technical approach for implementing orphaned well clean-up, as required by the Bipartisan Infrastructure Law.[3]

SPONSOR REMARKS

To provide important context for the workshop, Kimbra Davis, DOI, presented an overview of OWPO. She explained that on November 15, 2021, President Biden signed the Infrastructure Investment and Jobs Act (IIJA) into law, providing ~$4.7 billion over 9 years for states, tribes, and federal agencies to plug, remediate, and restore orphaned wells and well sites across the United States. OWPO was established in January 2023, she noted, to ensure that this investment is implemented effectively and efficiently to "transform a legacy of environmental pollution" across the United States into a "legacy of environmental stewardship." Of the $4.7 billion, ~$4.3 billion was set aside for wells on state and private lands, $250 million was allocated for activities on federal lands, and $150 million was reserved for activities on tribal lands.

Individual states, federal agencies, and tribes often have different procedures, standards, and requirements for plugging wells, and Davis noted that this workshop could provide an opportunity to discuss challenges and opportunities, share best practices, and learn from one another. She stressed that OWPO's goal is to understand the current landscape of well-plugging across the United States, why particular jurisdictions follow certain requirements and implement unique processes for well-plugging, and the merits of each state's approach in terms of environmental health and safety.

As of late 2021, states had identified almost 130,000 orphaned wells on state and private lands, and another 16,000 orphaned wells have been identified on federal lands—and both numbers are expected to continue to increase as more are identified. Davis suggested that sustained collaborative efforts among federal, state, and local officials; academia; nonprofits; industry; and the public to address the orphaned well problem are essential.

Davis indicated that the program as of March 31, 2024, had plugged more than 7,700 orphaned wells on state and private lands and as of July 18, 2024, awarded $960 million to 25 states. Additionally, $145 million was distributed to five federal land management agencies—Bureau of Land Management, National Park Service, Fish and Wildlife Service, Forest Service, and Bureau of Safety and Environmental

---

[3] The Bipartisan Infrastructure Law is also referred to as the Infrastructure Investment and Jobs Act.

Enforcement—that plan to plug nearly 600 orphaned wells on public lands and waters. (As of July 18, 2024, 190 of these wells had been plugged.) Furthermore, $40 million has been awarded to tribes to begin plugging wells later in 2024. In total, nearly 8,000 orphaned wells have been plugged with IIJA funds. She added that to complete this work, numerous jobs were created in both urban and rural areas across the United States, where millions of people live near wells (see Figure 1-1) that were posing serious health and safety risks.

Davis underscored that if orphaned wells are not plugged, methane can leak into the atmosphere and exacerbate climate change. Local air pollution issues also can arise if other hydrocarbons are leaking from these wells. Orphaned wells can prevent communities from recognizing the full economic potential of their land, and she noted that implementation of the Orphaned Wells Program thus presents both an "enormous challenge and a transformative opportunity."

ORGANIZATION OF THIS PROCEEDINGS

The chapters that follow summarize the presentations and discussions among workshop speakers and participants. Chapter 2 offers perspectives on the costs, challenges, and benefits of plugging orphaned and abandoned wells, including a discussion of various states' well-plugging and abandonment standards and procedures.[4] Chapter 3 describes how wellbore integrity and subsurface conditions are evaluated for well-plugging prioritization. Chapter 4 considers wellbore procedures and best practices, and Chapter 5 focuses on issues related to environmental risks and monitoring. Chapter 6 explores opportunities for and experiences with remediation, reclamation, and restoration. Finally, Chapter 7 highlights advances in plugging and abandonment for idled wells, and Chapter 8 presents examples of key workshop themes.

---

[4] The National Academies commissioned a white paper that compiles state practices and discusses statutory and regulatory standards, methods, and design of plugging plans and requirements for well-plugging activities. This white paper is available at https://nap.nationalacademies.org/resource/28035/White_Paper_Orphaned_Wells_Workshop_Proceedings.pdf.

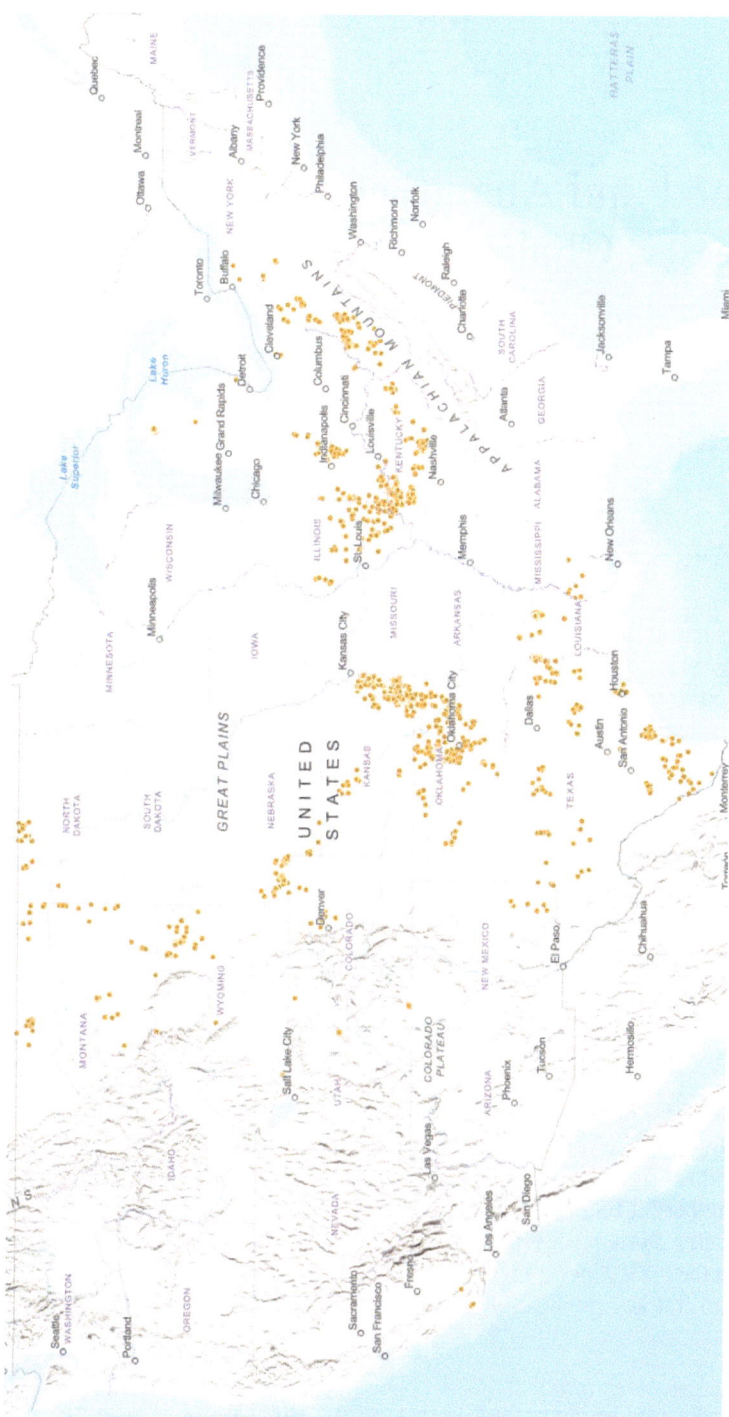

FIGURE 1-1 Locations of the more than 6,100 orphaned wells that were plugged on state, private, and federal lands between October 1, 2022, and September 30, 2023.
SOURCE: DOI Orphaned Wells Program Office, 2024.

2

Orphaned and Abandoned Well-Plugging: Costs, Challenges, and Benefits

INTRODUCTION

The first session of the workshop, moderated by workshop planning committee member James Slutz, National Petroleum Council, presented a high-level overview of issues related to plugging orphaned and abandoned wells. Speakers were asked to highlight the history of well-plugging and associated regulations and to explore the scope and scale of current well-plugging challenges.

Slutz emphasized that given the lengthy and complex history of production and regulation in the United States, managing its orphaned wells is difficult. He highlighted the guidance of *A Study of Conservation of Oil and Gas* (IOGCC, 1964), which noted that because of differences in geology, production, marketing, and economics that demand flexibility at the local level, "uniformity of conservation regulations among the states does not exist" and would likely be an unrealistic expectation.

HISTORIC AND CURRENT WELL-PLUGGING EFFORTS ACROSS THE STATES

Lori Wrotenbery, Interstate Oil and Gas Compact Commission (IOGCC), provided a brief overview of state efforts to manage orphaned wells over the past several decades. She used examples from a few specific states, including the Railroad Commission of Texas's drilling permit fee, New Mexico's data management system, and Oklahoma's Global Positioning System for reporting current locations of all known wells. Wrotenbery noted that the IOGCC works with all 38 oil- and gas-producing states.[1] The IOGCC has released several key publications, including a 2019 report titled *Idle and Orphan*

---

[1] This tally includes states that historically produced or are currently producing oil and/or gas.

Oil and Gas Wells: State and Provincial Regulatory Strategies.[2] She indicated that this report informed efforts to provide funding for states to plug documented orphaned wells when the oil and gas industry experienced decreased demand and pricing issues during the COVID-19 pandemic. The 2019 report was updated with new data in 2021,[3] and the IOGCC released a supplementary report in 2023[4] with information on states' prioritization systems for plugging orphaned wells. In 2024, a supplementary report[5] on orphaned well-plugging and site restoration was released, based on 3 years of survey data collected from individual states. She said that 141,959 orphaned wells were documented as of December 31, 2023 (with 29 states reporting), versus only 49,743 in 1992. States continue to improve their databases and examine historical records in order to refine their inventories, as states estimate that 250,000–740,000 wells remain undocumented.

Wrotenbery then shared other data from the 2024 supplementary report on the number of orphaned wells that were plugged in these 29 states. From 2021 through 2023, a total of 70,482 wells were plugged. States plugged more than 15,000 of those wells—nearly half of which were plugged using federal funds—and responsible operators plugged the remaining ~55,000 wells. The 2024 supplementary report also includes data on the costs associated with well-plugging, which vary widely within and across states depending on well depth, condition, location, and accessibility. Average state plugging costs ranged from $3,664 per well to $343,750 per well; across all 27 states that reported cost data, the average cost to plug a well in 2023 was $41,139. Although the states' costs vary significantly, she mentioned that costs overall have increased over the past 6 years and believes this is due to inflationary factors, and that competition for contractors has increased.

ORPHANED WELLS: LOCATIONS, ATTRIBUTES, AND COSTS

Adam Peltz, Environmental Defense Fund (EDF), noted a gap in the data collected on the locations of orphaned wells both because some states do not count orphaned *federal* wells in their totals and because an unknown number of undocumented wells remains. He indicated that of the 125,000[6] documented orphaned wells that had been counted as of November 2021 across 30 states, many are concentrated in Appalachia, followed by Kentucky and Illinois, south central United States (including Missouri, Kansas, Oklahoma, Texas, and Louisiana), the Rocky Mountains, and urban southern California (see Figure 2-1). He noted that 18 million Americans live within 1 mile of an active well, and 14 million Americans live within 1 mile of a documented orphaned well of those counted

---

[2] This report is available at https://oklahoma.gov/content/dam/ok/en/iogcc/documents/publications/2020_03_04_updated_idle_and_orphan_oil_and_gas_wells_report.pdf.

[3] This report is available at https://oklahoma.gov/content/dam/ok/en/iogcc/documents/publications/iogcc_idle_and_orphan_wells_2021_final_web.pdf.

[4] This report is available at https://oklahoma.gov/content/dam/ok/en/iogcc/documents/publications/prioritization_report_7.10.23.pdf.

[5] This report is available at https://oklahoma.gov/content/dam/ok/en/iogcc/documents/Idle%20and%20Orphan%20Wells.pdf.

[6] As mentioned by Wrotenbery, this number increased to 141,959 by the end of 2023.

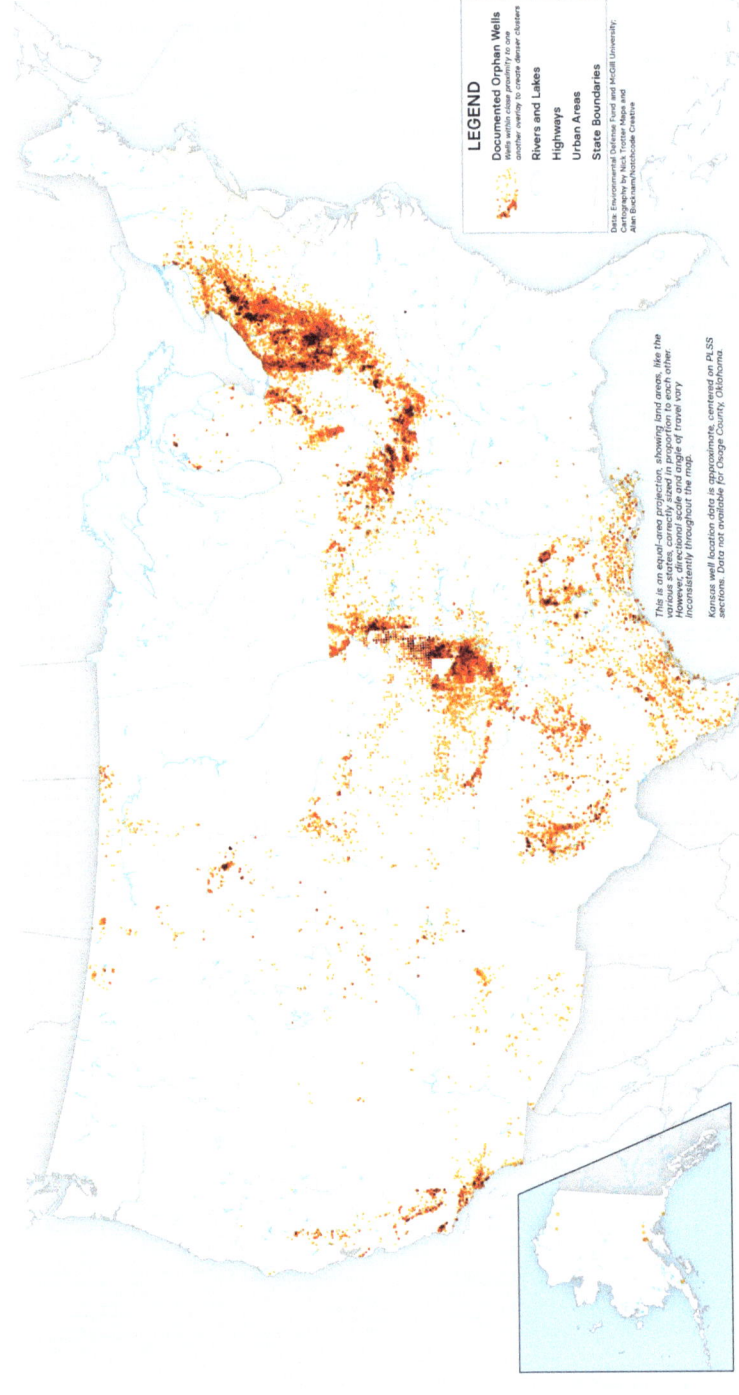

FIGURE 2-1 Documented orphaned wells in the United States as of November 2021.
SOURCE: Environmental Defense Fund, 2021.

by late 2021 (Boutot et al., 2022); furthermore, an overlap exists between locations with high concentrations of active wells and marginalized communities.

Peltz highlighted the increased efforts of several states to plug orphaned wells with Infrastructure Investment and Jobs Act (IIJA) funds—especially Kansas, Texas, Louisiana, and Kentucky—and he estimated that the total number of wells plugged in the United States with these funds is now approaching 10,000. He echoed Wrotenbery's comment that the estimated costs to manage orphaned wells vary considerably within and across states, especially because some calculations only include plugging costs and not remediation costs. For the most part, states with straightforward topography and geology and shallow wells have lower costs. Although well-plugging can be costly in some locations, he noted the associated benefit of job creation in the oilfield service sector. For example, in Louisiana, many thousands of job-years could be realized from plugging the state's ~17,000 idled wells.

Turning to known attributes of oil and gas wells, Peltz referred to a study of 82,000 documented orphaned wells, which found that much is known about well type (e.g., oil vs. gas), with data available for 83% of those documented orphaned wells. Less is known about well depth, with related data available for 50% of the wells, which leads to challenges in estimating accurate costs to plug wells. Even less is known about last production date (data are available for only 16% of the wells), which is a key factor in determining how to plug a well (Boutot et al., 2022).

Drawing on data from another study of the same 82,000 documented orphaned wells (Kang et al., 2023), Peltz described environmental and locational attributes of documented orphaned wells. He indicated that although knowledge on the relationship between orphaned wells and groundwater quality and contamination is limited, protecting groundwater is a primary reason to plug orphaned wells, especially given how many of these wells are located within close proximity to domestic groundwater wells. For example, a study on behalf of the Ground Water Protection Council[7] (Kell, 2011) analyzed records of water contamination related to oil and gas in Texas and Ohio and found that nearly 20% of the contamination cases could be traced to an orphaned or abandoned hydrocarbon well. He pointed out that wells that have been leaking for decades can cause significant contamination issues, which are especially expensive to address. Concerns also arise related to methane emissions and to air quality issues from other pollutants.

The study by Kang and colleagues (2023) also explored potential beneficial uses for documented orphaned wells, such as for energy storage and geothermal heat. Furthermore, sites of documented orphaned wells could be repurposed for wind and solar activities. However, Peltz noted that carbon sequestration in areas with high concentrations of documented orphaned wells is problematic—and even more dangerous in areas with an unknown number of undocumented wells.

In his closing remarks, Peltz summarized the results of an analysis conducted across nine jurisdictions of the written rules[8] associated with the Ground Water Protec-

---

[7] Kell, S. (2011). State Oil and Gas Agency Groundwater Investigations and their Role in Advancing Regulatory Reforms, A Two-State Review: Ohio and Texas. Oklahoma City, OK: Ground Water Protection Council.

[8] Peltz noted that states have additional guidance that is not captured in these rules.

tion Council's "Regulatory Elements for Well Integrity".[9] Sixty (of 175) issues raised by these regulatory elements were found to need improvement on, for example, identification of water intervals that need cement protections; protection of other natural resources; establishment of standards for cement quality, slurry prep and placement, and mix water quality; expansion of pre-plugging wellbore conditioning standards; and determinations for testing the plugs. He emphasized that plugging standards can always be amended as technology advances and highlighted the importance of identifying gaps and effective approaches as well as learning from one another's experiences.

RESEARCH AND DEVELOPMENT PROGRAM FOR UNDOCUMENTED ORPHANED WELLS

David Alleman, Department of Energy (DOE), described the Office of Fossil Energy and Carbon Management's (FECM's) ongoing methane mitigation research program, which aims to quantify and reduce emissions across the natural gas supply chain. Part of that program focuses on developing tools, technologies, and processes to efficiently identify and characterize undocumented orphaned wells for plugging and abandonment prioritization.

Alleman highlighted current initiatives with which FECM is involved. First, he mentioned a funding opportunity announcement for the Environmental Protection Agency's (EPA's) Methane Emissions Reduction Program. A total budget of $1.3 billion is available to plug marginal wells and to deploy technologies to help small operators reduce emissions. When EPA's new regulations are implemented, he indicated that these small operators will be better prepared to comply. Second, while the Department of the Interior's (DOI's) Orphaned Wells Program was given a budget of $4.7 billion as a result of the IIJA, DOE was awarded $30 million over 5 years specifically for its Undocumented Orphaned Wells Program. An additional $4 million from annual appropriations was also dedicated to this effort. With these funds, DOE is working with states, tribes, and federal land management agencies as well as with the IOGCC to develop technologies and techniques to identify and characterize orphaned wells that are not in the regulatory inventory. This program has nine priority areas: methane detection and quantification, well identification, sensor fusion and data integration with machine learning, well characterization, integration and best practices, data management, records data extraction, well database creation, and use of field teams. Furthermore, he referenced a Notice of Intent released in July 2024 for an upcoming funding opportunity announcement,[10] with a focus on how states can address issues related to advanced remediation, wellbore characterization, and long-term monitoring of undocumented orphaned wells.

Alleman then turned to a discussion of the challenges of this work, particularly how to find, characterize, and plug undocumented wells. He emphasized that although

---

[9] This analysis can be found at https://www.gwpc.org/wp-content/uploads/2021/03/Well_Integrity_Elements_Revised_1_19_2021_002.pdf.

[10] This Notice of Intent is available at https://netl.doe.gov/node/13935.

no "silver bullet" exists to identify undocumented orphaned wells, useful technologies include magnetic surveys, aerial and satellite photographs, LiDAR, methane measurements, and state historical records. Research continues to be conducted to improve sensors and, in particular, data processing. Drone technology currently enables much of the identification work, and large drones provide more payload and more capacity. However, he pointed out that land access challenges arise with the use of drones, which require permits or Federal Aviation Administration "beyond visual line of sight" waivers.

After the wells have been identified, they are characterized. DOE is developing technologies to evaluate well condition without entering the wells. His team is also working to identify a robust set of sensors that can locate orphaned wells efficiently at scale—for instance, wells that do not appear with magnetometry could appear in overhead imagery. However, he noted that significant effort is required to synthesize the data collected from multiple sources to try to locate undocumented wells. He added that states with many documented orphaned wells are likely to have a high number of undocumented orphaned wells; furthermore, states with the longest production history are likely to have the most undocumented wells. He estimated that 800,000 undocumented orphaned wells remain in the United States.

Once orphaned wells are documented, Alleman explained that they can be prioritized for plugging and abandonment. Plugging is difficult in cases in which little is known about the well, and costs of bringing equipment to the well to characterize it are high; therefore, he encouraged learning as much about the well as possible ahead of time. An additional challenge arises when plugging wells in ecologically sensitive areas that affect the infrastructure that can be built to reach the wells and the materials that can be used to plug them. He underscored that tradeoffs between environmental benefit and damage have to be weighed carefully.

COMMENTARY FROM STATE OIL AND GAS LEADERS

As the first session of the workshop continued, Slutz invited five state leaders to offer their perspectives on how the topics of the opening presentations relate to issues faced by their respective states.

Tom Kropatsch, Wyoming Oil and Gas Conservation Commission, emphasized that Wyoming has been plugging orphaned wells for decades and adheres to the same standards and regulations that the oil and gas industry uses to ensure protection of both the environment and public health and safety. When plugging an orphaned well, he pointed out the importance of considering a property owner's current and future surface uses as well as the varied situations that states and federal agencies encounter. He stressed, however, that anticipating everything that could happen when plugging a well is impossible, and Wyoming's regulations are thus designed to allow flexibility. This perspective may have influenced DOI's decision not to create a standardized process across the United States for plugging wells.

Eric Vendel, Ohio Department of Natural Resources, indicated that because many of the current problems with Ohio's wells were created between the mid-19th and mid-

20th centuries, various histories and vintages of wells are considered when planning to plug wells. The Ohio Department of Natural Resources established its orphaned well program in 1977, and $130 million of state funds has been used to plug wells. He underscored that developing the most applicable funding mechanisms is key to continuing this work; for example, in 2012, when the shale industry began drilling wells, Ohio law was restructured to base funding on a severance tax, which increased the budget to allow for more plugging. He also discussed the value of researching the longevity of cement and other plugging materials—although cements used in Ohio (i.e., Class L and Type 1L) have been tested in the laboratory with acceptable results, different situations and challenges arise in the field.

Danny Sorrells, Railroad Commission of Texas (RRC), noted that Texas has been plugging and re-plugging wells for more than 100 years, with 10 state rigs and 11 federal rigs currently active across the state. Over the past 42 years, ~48,000 wells have been plugged using state funds (~$500 million). In 2023, 762 wells were plugged with IIJA funds, and more than 1,000 wells were plugged with state funds. Like Kropatsch, he expressed his support for DOI's decision not to pursue a standard approach to well-plugging across the nation. For instance, because of the varied geology across Texas, different approaches to plugging wells are used across its 10 districts. Expanding on Vendel's suggestion, he said that current cementing practices are outdated and likely not the most effective. He encouraged the development of new cementing technologies—for example, alternate materials such as geopolymers could be studied further.

Don Hegburg, Pennsylvania Department of Environmental Protection, explained that Pennsylvania has a 165-year history of oil and gas development. Only 27,000 orphaned wells are documented out of the 330,000 wells that, according to historical records, have been drilled, and more than 200,000 of these wells were drilled prior to the 1950s. Thus, he indicated that Pennsylvania would be using its IIJA funds to locate and begin to address undocumented wells. Furthermore, Pennsylvania will be integrating drones and other technologies into the process for locating undocumented orphaned wells and will be developing protocols in consideration of pilot-scale surveys that will be conducted by the Environmental Defense Fund in Pennsylvania later this year. He observed that ~3,400 wells have been plugged over 40 years, and Pennsylvania encounters many of the same problems as other states—for example, geological and geochemical issues that deteriorate wells and casings. Pennsylvania also has thousands of unconventional wells and a history of deep mining, both of which further complicate the plugging process. He acknowledged that although Pennsylvania has robust plugging standards, they could be improved with new approaches that extend plug longevity (e.g., metallurgical technologies that use bismuth).

Bryan McLellan, Alaska Oil and Gas Conservation Commission, remarked that Alaska has several unique issues related to well-plugging. Accessing and plugging old, remote orphaned wells might cost millions of dollars and be limited to certain seasons (e.g., driving across the tundra in the summer is impossible, and an ice road is required in the winter). Often helicopters are used to reach well sites, many of which are discovered based on 100-year-old hand-drawn maps. Thus, he expressed his interest in collaborating with DOE to use magnetics to locate wells more easily in remote areas.

He added that Alaska did not have a state-run orphaned well program prior to the IIJA era. Therefore, the initial IIJA grant of $25 million was used to establish the program, with assistance from other state agencies to begin contracting large-scale operations; to locate and investigate known orphaned wells; and to identify additional orphaned wells. Thus far, ~50 orphaned wells are known or suspected across Alaska—a number lower than that of many other states because Alaska has a newer oil and gas program, and the relatively higher cost of drilling wells is more compatible to working with major oil and gas companies who plug their own wells. The rest of the initial funding likely will be used to plug wells within 2 miles of existing road systems, and the $29.3 million IIJA formula grant is earmarked to plug Katalla Oilfield; however, he said that additional funding will be needed, given how difficult that site is to access.

OPEN DISCUSSION

Slutz moderated a discussion among the workshop speakers and participants. An online participant asked the speakers to provide a definition of "orphaned well." Wrotenbery first indicated that "orphaned wells" and "abandoned wells" are not necessarily the same. EPA broadly defines "abandoned wells" to include already-plugged wells, orphaned wells, and wells that have responsible operators but that are inactive for a certain length of time. The IOGCC defines an "orphaned well" as one that is inactive and does not have a viable operator that is responsible for plugging and clean-up. However, to complicate matters further, she noted that terms and their definitions vary across states.

Another online participant asked how states prioritize wells for plugging, especially when funds are limited. Wrotenbery replied that the IOGCC's 2023 supplemental report[11] provides information on states' prioritization systems, including the factors they consider and the scoring they use. She added that the IIJA's orphaned well provisions defer to each individual state's priority ranking system—and many of those priorities are first and foremost based on public safety, followed by groundwater and surface water protection.

Dwayne Purvis, Purvis Energy Advisors, inquired about the possibility of setting up a long-term monitor on wells that are not an immediate priority. Alleman responded that such a decision would be made by the states; however, DOE is searching for low-cost continuous monitoring approaches that could both identify problematic wells and measure the effectiveness of recently plugged wells.

An online participant asked about the relationship between seismic events and well failure. Workshop planning committee member Mary Kang, McGill University, explained that an analysis of data collected in British Columbia found a relationship between seismic events and areas where hydraulic fracturing is conducted. The study also measured methane emissions from plugged and unplugged wells; the study found a statistically significant relationship with higher methane emissions observed at plugged

---

[11] This report is available at https://oklahoma.gov/content/dam/ok/en/iogcc/documents/publications/prioritization_report_7.10.23.pdf.

wells in areas experiencing more frequent earthquakes, as well as areas near earthquake occurrences. Eric van Oort, University of Texas at Austin, described a machine learning study for wells in New Mexico that found that the proximity to seismic events was a factor in well leaks—even for plugged wells. Given these correlations, Kang and van Oort suggested that further research be conducted.

Another online participant inquired about other extractive industry unknowns. Peltz said that at least three unknowns exist—the number of undocumented orphaned wells in the United States, the number of inactive unplugged wells in the world, and plug failure rates over time. Dan Arthur, ALL Consulting, highlighted work on projects in Venezuela, Ecuador, India, China, and Albania and encouraged sharing learnings with those working on the same types of issues in other countries.

An online participant posed a question about repurposing wells. Peltz replied that DOE's geothermal unit is considering strategies to repurpose wells (with geothermal heat having more promise than geothermal power because the latter requires larger well diameters than those of many end-of-life wells). For example, in some end-of-life wells in Oklahoma, water is being injected into two wells and produced in two other wells to generate heat for local elementary and high schools. Additionally, wellbores could be used for energy storage. He mentioned that the Abandoned Well Remediation Research and Development Act would, if passed, direct $160 million to DOE to discover undocumented orphaned wells, examine plugging technologies (e.g., to address issues with cement), and consider beneficial uses of end-of-life wells. He added that repurposing wells and sites could generate new revenue sources to plug other wells. Alleman noted that although repurposing wells is desirable, many wells—especially older, undocumented wells with compromised wellbores—cannot be repurposed.

Michael Hickey, Colorado Energy and Carbon Management Commission, reflected on the funding sources for remediation. He suggested consideration of carbon credits for impacted soil. He said that industry partners are plugging wells at no cost to the state, but they do not touch impacted soils; if carbon credits were available for impacted soils, the atmosphere and the groundwater could be better protected. Arthur highlighted that many idled and marginal wells that have no operator might become orphaned but are not yet defined as such by the states. He proposed that this issue be addressed via the private-sector carbon market. He added that many wells are located around schools, businesses, and homes and in waterways where people are getting their drinking water, but the potential impacts are not yet understood. Complicating this issue further is that industry plans to drill 20,000 geothermal wells a year until 2050; although this would significantly increase the energy supply, he pointed out that those wells also would likely need to be plugged. Peltz said that those geothermal wells could be required to be bonded before being drilled, which could eliminate concerns about future orphaning. Wrotenbery noted that policy discussions continue on the specific issue of carbon credits. The use of the carbon credit market currently is prohibited for wells plugged or surfaces remediated with federal funds; however, states are asking DOI to reconsider that position because carbon credit participation could extend funds and allow states to plug more wells. Peltz posed several key questions about the relationship between plugging orphaned wells and the use of the carbon credit market: How do you estimate how much methane is coming from wells? How long will the crediting period extend? What if you plug one well and the methane migrates to other wells? He emphasized that uncertainties remain about how to have a high-integrity carbon market.

Slutz noted that this discussion could inform the upcoming National Academies of Sciences, Engineering, and Medicine consensus study on orphaned and abandoned wells. Peltz replied that the next steps could be to facilitate conversation between states and federal agencies on best practices and lessons learned, and to make a list of plugging technologies that can be further researched. Wrotenbery encouraged increased conversation with states to better understand their needs; she highlighted an IOGCC committee that identifies technical issues of greatest concern to regulators each year and then creates related webinars and research papers. Alleman also spotlighted the states' wealth of knowledge and invited increased collaboration with DOE. In particular, he urged states to volunteer sites where the national laboratories could test new technologies for locating wells.

COMPILATION OF STATE STANDARDS AND PROCEDURES FOR PLUGGING AND ABANDONING WELLS

Rick Simmers, formerly of Ohio Department of Natural Resources, and co-authors were invited by the National Academies to submit a white paper compiling shared practices, statutory and regulatory standards, methods, policies and procedures, and designs of plans for well-plugging activities across state oil and gas regulatory programs.[12] To gather data for this white paper, the authors disseminated a questionnaire for states about the elements of their respective plugging processes (e.g., identifying, locating, and characterizing wells and selecting materials). Responses to the questionnaire helped the authors to characterize the various methods used by program managers to address problems associated with orphaned wells.

Before discussing specific data from the questionnaire, Simmers provided a brief overview of drilling methods used in the United States over the past 150 years. The earliest method of drilling used a spring pole—essentially beating a hole in the ground to create a well. The next method relied on a mobile rig such as a derrick on a wagon. He explained that neither method reached significant depths, and both had limited capabilities. In Long Beach, California, where multiple rows of derricks were placed close together, when one well was plugged, the other wells would be affected and perhaps begin to flow. The cable tool rig was used to drill thousands of wells across the country. Particularly in the Appalachian Basin, some wells are still drilled using this method. More recent methods include rotary rigs that are modified to drill shallow portions of wells (and thus offer a less expensive drilling approach). He indicated that any combination of these more modern rigs could be used to plug a well.

Simmers noted that GPS data are often used to document orphaned wells. He stressed that advanced technology to help locate wells could be valuable, especially

---

[12] This white paper is available at https://nap.nationalacademies.org/resource/28035/White_Paper_Orphaned_Wells_Workshop_Proceedings.pdf.

for complex settings like wooded areas or those with dense vegetation. Once wells are located, program managers face many challenges. For example, some wells might still have equipment in them; in other cases, program managers might be confronted with open annuli or a repaired well. When records of the original drilling do not exist, program managers may have a more difficult time developing plans to best plug the well. Furthermore, he continued, orphaned wells that are found in waterways are difficult to access and to plug—but it is important to prevent leakage from affecting the waterways. When wells are found in wetlands, program managers first determine if the wells were constructed in a way to protect groundwater, which may be defined differently by state. Potentially even more challenging to plug are orphaned wells located underneath infrastructure such as schools. To prevent this issue in the future, he suggested that GPS coordinates of newly located wells be shared throughout communities so that new structures are not constructed over them.

Returning to a discussion of the questionnaire, Simmers highlighted a few of the questions that state representatives were asked:

- Do you have a list of approved plugging materials?
- Do you define by regulation which plugging material should be used and which alternate materials are allowed?
- How are plug placement and volume defined by regulation?
- Are other items allowed to be used as plugs, as defined by regulation?
- What are the requirements around plug spacers and plug placement methods?

Simmers described some of the responses to the question about approved plugging materials. Alabama and Wyoming require the use of any American Petroleum Institute (API) cements and additives. Arkansas specifically requires the use of API Class A and Class H materials; however, Class A materials are becoming scarcer. Colorado focuses on specific performance criteria instead, requiring cement to reach 300 psi after 24 hours and 800 psi after 72 hours at 95 degrees. Oklahoma and Pennsylvania are particularly interested in performance standards such as compressive strength and permeability for plugging materials. He mentioned that all states that responded to the questionnaire (from here forward referred to as "respondents") considered Portland-based cement as a default plugging material; furthermore, most of the states' programs have both prescriptive and performance-based measures for the plugging material.

Simmers explained that all respondents have processes in place to approve the use of alternate plugging materials and equipment. These materials and equipment might be part of either a primary or secondary plugging plan. Generally, respondents recognize instances in which a primary plan to plug a well will not work, and all respondents have a provision to consider an alternate plan, alternate materials, or both. However, he stressed that alternate plans typically undergo a rigorous review and approval process to assure protection of people and resources.

Once materials have been selected and approved, Simmers noted that the next step is placing the materials in the well to plug it. Some states' regulations identify the isolation of production and target zones (e.g., a mineral resource of future value) as a key consideration for plug placement methods, and all respondents' regulations require the isolation of "protected water." Some respondents define specifically how to place plugs—for example, Arkansas allows the use of pressurized cementing for plug placement or the use of a dump bailer (i.e., gravity cementing). States that do not define how to place plugs choose their methods based on individual well constructions and plug requirements. Pressurized systems are the most popular choice, he explained, when well construction allows for them.

Summarizing general approaches to plug placement, Simmers mentioned that, for a bottom plug, some states require cement or plugging material to be placed below the production zone (and below perforations, if they exist) and additional cement to be placed above the production zone. When bridge plugs—a mechanical device that acts as a temporary barrier—are allowed, some respondents' regulations require that a certain amount of cement be placed on top of them and that they serve as intermediate plugs, while others require that they are used only as temporary plugs. Lastly, when surface plugs are placed to protect water, respondents have very different approaches to determine required depths.

Simmers said that regulatory witnessing also varies significantly by state—for example, from 8 hours of advanced notification of plugging in Arkansas to 5 business days of advanced notification of plugging in Missouri. Furthermore, Arkansas can require a plug to be drilled out if notice of plugging was not received in the required timeframe. Work that might be observed by regulators includes tagging of plugs, monitoring of annular spaces and fluid levels, or placing of plugs. Although no regulatory program can witness every part of every plugging job, operators know that inspections can occur at any time. Additionally, all respondent regulations require the owner/operator or plugging contractor to submit reports (within ~30–60 days) detailing how the well was plugged, who was there, and if anything was left in the well that was not used for plugging; service company record attachments might also be required. He reiterated that all of these steps help to ensure that a well is plugged correctly and no longer contributes to human health and safety and environmental problems, and that land is restored for future use.

Discussion of the Compilation of State Standards and Procedures for Plugging and Abandoning Wells

An online participant asked if any states require an evaluation log of cement outside the casing prior to plugging within the casing. Simmers replied that many states require evaluation of proper backside isolation if a casing is left in a well. Furthermore, many states require wells to be plugged in a static condition (i.e., no flow), with isolation to the formation of origin.

In terms of regulations and approval of plugging plans, Arthur wondered if states and/or the federal government are considering how to account for the fact that the zones deemed worthy of protection may change in the future. Simmers said that some states have defined target zones, and states could benefit from periodically evaluating their targets and what they wish to protect (e.g., a carbon sequestration zone). He stressed that geology is a critical consideration.

Workshop planning committee member Nathan Meehan, Texas A&M University, asked if any states require long-term monitoring of methane emissions from plugged wells, reflecting that Alberta, Canada, requires testing wells for leaks 1 year after plugging. Simmers responded that after plugging, many states require the casing to be cut off down to a certain height (e.g., Ohio is 30 inches, some states require approximately 3–4 feet below grade), an identifying plate to be welded on top, and/or a monument (e.g., a pipe with identification information) to be installed. He added that Pennsylvania has long-term monitoring for mine-specific plugs, and in Ohio, some urban orphaned wells have permanent monitoring systems to detect gas. Kang posed a follow-up question about how often other wells in urban areas and areas with minable coal are monitored. Simmers noted that in Ohio, sampling would occur monthly up to 1 year; the regulator's monitoring would stop when no leakage was evident from the port, and the local fire department could continue monitoring at the request of the property owner.

Reflecting on a new EPA rule that would require a forward-looking infrared (FLIR) camera observation of a well after it has been plugged, Peltz wondered how long it would take for a problem (e.g., micro-annulus, gas flow) to become apparent to the surface-level camera. Simmers remarked that how a well was plugged and verified determines how long it would be monitored. He emphasized the importance of verification throughout the plugging process to ensure that plugs are placed properly and are not leaking. He added that as cement is pumped, it might pick up natural gas that will be detected by the FLIR camera 1 day after plugging; however, the well is not actually leaking, and gas bubbles will disappear in a few days. Jesse Frederick, WZI, mentioned that hydrogen could be present a few days after setting a surface plug but would not be captured by a FLIR camera. A workshop attendee explained that in Ohio, a well has to be left open for 72 hours after plugging to ensure that no additional plugging work may be necessary. When gas is observed during that timeframe, samples are collected and typically hydrogen is seen instead of methane. In those cases, the attendee stated that the problem usually self-ameliorates over time and does not prompt re-plugging. Meehan suggested looking at wells plugged 6 months ago or 1 year ago; if one-third of those wells have noticeable volumes of methane, it is possible that a serious problem exists.

3

Examples of Well-Plugging Prioritization Considerations: Evaluating Wellbore Integrity and Subsurface Conditions

INTRODUCTION

The second session of the workshop was moderated by workshop planning committee member Mileva Radonjic, Oklahoma State University. It focused on examples of strategies to prioritize well-plugging projects by evaluating wellbore condition (with consideration of integrity issues, historical records, and available data) and subsurface geology (including subsurface fluids surrounding the wells).

EXAMPLES OF PRIORITIZATION CONSIDERATIONS FOR MANAGING PERMANENT END-OF-LIFE SOLUTIONS FOR MARGINAL, IDLED, ORPHANED, AND OTHER WELLS

Dan Arthur, ALL Consulting, explained that understanding the varied risks of marginal, idled, orphaned, and other wells is critical for prioritization efforts. Prioritizing the plugging of more than 3.5 million idled and orphaned wells in the United States is complicated, he continued, owing to challenges related to the Permian Basin water crisis; fresh water shortages; historic plugged-well failures; produced water disposal; development of new resources (e.g., geothermal); and new, complex projects (e.g., carbon dioxide sequestration and hydrogen storage). He explained that understanding the interconnectivity of these and other issues (e.g., methane leaks, groundwater contamination, and oil seepage) might be one of the most significant challenges that the industry and regulators have faced. Furthermore, many discrepancies remain between the number of wells that have been inventoried and the number of wells that exist. For example, he said that in one area of Tulsa, Oklahoma, for every documented well, three to five undocumented wells exist. And within a 20-foot radius, one could expect to find one well but instead find five to six wells drilled to different depths right next to each other.

Given that the first commercial gas well was drilled in 1821 and the first commercial oil well was drilled in 1859, Arthur highlighted the importance of understanding the impacts of corrosion over time. He noted that learning about historical and pre-regulation environments, practices, and standards—and combining that historical knowledge with current perspectives—may help with understanding which wells to address first.

Exploring the prioritization process in more detail, Arthur observed that location is a key factor. He shared an anecdote from 1954 about a Tulsa home filling with gas from an unplugged abandoned well beneath it; additional unplugged wells were found beneath the local elementary school and nearby surface lots. Many oil and gas wells were drilled in now-populated urban areas, including economically disadvantaged communities, and many urban development areas encroached on highly developed oil and gas fields. For instance, in Kellyville, Oklahoma, 275 idled and orphaned wells—many of which are emitting methane—are within only 2 miles of four schools. In more populated areas, he underscored that the problem is even worse.

Arthur indicated that addressing and mitigating the impacts of methane leaks is a serious concern in the United States. He shared images captured by a forward-looking infrared camera from a well on his own property in Hectorville, Oklahoma, with methane leaking through the fittings and valves and migrating through the soil. In some applications, multiple methane flow meters of different ranges are required. He stressed that there is not necessarily a "one-shot fix all" approach, but new technologies are being developed that do volumetrics.

Arthur summarized that well-plugging prioritization is difficult owing to potential impacts to surface water or groundwater; well depth, age, and condition; threats to public health and safety; threats to the environment and wildlife; proximity to people and critical infrastructure; impacts to land and land use; soil impacts, leaks, and spills; and environmental justice issues. He underscored that well-plugging and site restoration is complicated and potentially dangerous work.

ORPHANED WELL-PLUGGING PRIORITIZATION IN WYOMING

Tom Kropatsch, Wyoming Oil and Gas Conservation Commission (WOGCC), provided a brief overview of oil and gas production in Wyoming. Since the first well was drilled in 1884, more than 118,000 wells have been drilled on federal, state, and private lands. Wyoming is ranked seventh overall in oil production and ninth overall in gas production in the United States. With ~60% of the mineral estate in Wyoming owned by the federal government, he noted that significant coordination occurs between state and federal agencies.

The WOGCC was created in 1951 and has operated an orphaned well program for several decades. Since 1997, this program has plugged more than 6,200 oil and gas wells on state and fee land. Since 2013, 500–600 wells have been plugged each year, and since the initial Infrastructure Investment and Jobs Act (IIJA) grant was awarded, 959 wells have been plugged. Kropatsch added that 335 more orphaned wells are under contract to be plugged by the end of 2024, and another 550 orphaned wells will need to be plugged using other funds.

Turning to a discussion of well prioritization, Kropatsch explained that requirements are set by state regulation:

> The Supervisor shall establish and maintain a well plugging schedule which prioritizes wells for plugging through an assessment of the well's potential to adversely impact public health [and] public safety [i.e., well proximity to public, which is less common in Wyoming], surface or ground waters [i.e., environmental impacts, which often can be prioritized first in more remote areas of the state], surface use [i.e., landowner loss of use or impact to operations], or other mineral resources. (055-3 Wyo. Code R. §§ 3-16(f))

He noted that prioritizing based on issues of well integrity involves understanding field and well history. For example, wells in the same field typically exhibit integrity issues at the same depths. Well construction, completions, geology, and fluids interact to contribute to these integrity issues. However, he pointed out that wells drilled later might not have the same cement and formation issues if knowledge from earlier wells was used to change drilling and completion. Understanding all of these factors is important in prioritizing plugging efforts, he continued, especially for wells that have had or may eventually need remedial work. In addition to understanding known issues and any changes made to a drilling program, he pointed to the importance of knowing the operator and how issues were handled, as well as whether routine mechanical integrity tests are being conducted.

Recognizing an area's geology is another key aspect of understanding well integrity and prioritizing wells for plugging, Kropatsch remarked. For example, salt zones in certain fields can cause casing to collapse, and sour gas zones can create safety issues for workers and the public as well as issues in the casing. He suggested that historical well records, production records, knowledge of nearby problem wells, and knowledge of the type of fluids in a wellbore all help to understand these well-integrity issues and to prioritize appropriately.

Kropatsch indicated that issues of well integrity impact plugging operations because equipment could get stuck in the well; cement squeezes may be necessary to remediate a failed casing; and obstructions, fill, and casing collapse could prevent access at the right depth. Such issues could increase the cost and time required to complete a project as well as decrease the effectiveness of a plug. Therefore, he emphasized the usefulness of understanding the impact of integrity issues on the environment, public health and safety, downhole issues, and plugging conditions prior to prioritization. Wildlife, weather, timing, landowner access restrictions, and surface issues are also considered. Wyoming then uses a ranking system to prioritize plugging wells with the highest potential impacts. He suggested that lower-priority wells of similar type, geography, and plugging requirements could be added to a plugging project for a higher-priority well to help ensure efficient use of funds.

ORPHANED AND ABANDONED
WELL-PLUGGING IN PENNSYLVANIA

Don Hegburg, Pennsylvania Department of Environmental Protection (DEP), reiterated that to begin prioritizing wells for plugging, it is important to understand the state's history of drilling and associated regulations. In Pennsylvania, the first oil well was drilled in 1859. Wooden plugs were used in the early 20th century before modern standards for plugging with cement emerged in the 1950s; currently, Pennsylvania has cement plugging requirements through oil- and gas-producing and coal extraction zones but does not have standards for protection/isolation of other formations that might be useful for mineral extraction or storage/disposal of fluids and gases.

Hegburg noted that DEP's current inventory includes ~173,000 conventional wells, and its well-plugging program was established in 1985 to plug oil and gas wells without an identifiable responsible party. As mentioned in the first session of the workshop, DEP has documented more than 27,000 abandoned wells and plugged ~3,400 wells over the past 40 years.

Before IIJA funding was available, Hegburg explained that an average of 21 wells were plugged and about $1 million was spent per year in Pennsylvania. Now, the IIJA funding will provide ~$400 million to Pennsylvania to plug orphaned and abandoned wells through 2031, and the budget is expected to be ~$50 million per year for the next 4 years. While the average plugging cost under state contracts from 2014 until 2023 was ~$50,000 per well, he said that since the IIJA contract was issued, that cost has increased to ~$106,000 per well. With the initial IIJA grant, Pennsylvania has plugged 199 wells and plans to plug 26 more wells by the end of 2024.

Hegburg then moved to a discussion about downhole challenges that affect well-plugging. For example, oil and gas wells vary from single to multistring casing designs; some wells are more than 100 years old, without reliable records; and well integrity is often compromised, which causes problems with retrieving casing, reaching total depth, and placing cement plugs. In one example, he cited a pH of 1 to 2 in groundwater in some areas, which would completely deteriorate casing. Several surface challenges also exist, and well access can be a significant cost.

Historically, Hegburg continued, well-plugging priority was determined after conducting site investigations that included assigning a numeric score based on a well's risk. The primary risks of abandoned wells in Pennsylvania include stray gas migration into homes and private water wells, fugitive methane emissions or hydrogen sulfide releases into the atmosphere, liquid releases into surface water and groundwater resources, casing deterioration through interaction with acidic mine-influenced water, and well communication events during new well completions.

With the IIJA funds, however, Hegburg indicated that the prioritization process for plugging wells is changing. Instead of prioritizing only emergency well-plugging activities, additional funds may allow for the prioritization of methane monitoring and quantification; environmental remediation; undocumented well (estimated around 300,000 in Pennsylvania) discovery with record review, georeferencing, drone surveys, and field verification; environmental justice outreach, including public meetings; and workforce development. Pennsylvania's new priority scoring system was developed using ArcGIS Survey123.[1] The scoring system includes 8–10 categories and 20–30 subcategories—for example, proximity to sensitive receptors, buildings, water supply, or sensitive habitats. Inspectors can enter data into the Survey123 app, which the central office can then access and begin to score to inform priority rankings. He stressed that Survey123 is a transparent and justifiable way to support decision-making, and DEP's public-facing website[2] tracks project progress and costs and provides emissions data (see Figure 3-1).

RAILROAD COMMISSION OF TEXAS: STANDARD PRACTICES FOR PLUGGING ORPHANED AND ABANDONED WELLS

Danny Sorrells, Railroad Commission of Texas (RRC), noted that the RRC's well prioritization process has four categories—well completion; wellbore conditions; well location with respect to sensitive areas; and unique environmental, safety, or economic concerns—with subcategories and scores for each. When determining how to score wellbore conditions, for example, he said that an ideal well is one from which the pump, rods, and tubing can be pulled and for which the casing has integrity—a rare find in Texas. Unforeseeable issues include pumps that are stuck downhole, collapsed casing, parted rods or tubing, holes in the casing, and debris in the hole. When determining how to score wells in sensitive areas, one primary concern may be proximity to the public.

Sorrells explained that the RRC's scoring within these four categories results in a ranking of 1, 2H, 2, 3, or 4 for prioritization; a score of 1, which represents a leaking well, is automatically prioritized for plugging. Inspectors view wells periodically to evaluate whether scores and rankings should be updated.

Sorrells indicated that standard plugging requirements in Texas are based on the Texas Administrative Code (TAC);[3] for example, "Wells shall be plugged to ensure that all formations bearing usable quality water, oil, gas, or geothermal resources are protected" (16 TAC §3.14(d)(1)), and "cement plugs shall be set to isolate each productive horizon and usable quality water strata" (16 TAC §3.14(d)(2)). He explained that the tagged plug is the most important plug in the well; the deepest water plug is required to be tagged by tubing, drill pipe, or approval by the district director for wireline (16 TAC §3.14(d)(2)). Furthermore, "plugs shall be placed by circulation or squeeze method through tubing or drill pipe" (16 TAC §3.14(d)(3)), and "all cement for plugging shall be an approved API [American Petroleum Institute] oil well cement without volume extenders" (16 TAC §3.14(d)(4)). Although exceptions may be granted in some cases, the regulations also state that "mud-laden fluid of at least 9 1/2 pounds per gallon with a minimum funnel viscosity of 40 seconds shall be placed in all portions of the well not

---

[1] For more information about this tool, see https://www.esri.com/en-us/arcgis/products/arcgis-survey123/overview?srsltid=AfmBOorht1YqAGsGtx8ZdFNs4T_IG9nStnFGZT4CgbE42iU6FRbPxlpZ.

[2] DEP's project tracker is available at https://padep-1.maps.arcgis.com/apps/instant/portfolio/index.html?appid=064e373125c34182b2e132dd50d7c619.

[3] To view the TAC, see https://texreg.sos.state.tx.us/public/readtac$ext.TacPage?sl=R&app=9&p_dir=&p_rloc=&p_tloc=&p_ploc=&pg=1&p_tac=&ti=16&pt=1&ch=3&rl=14.

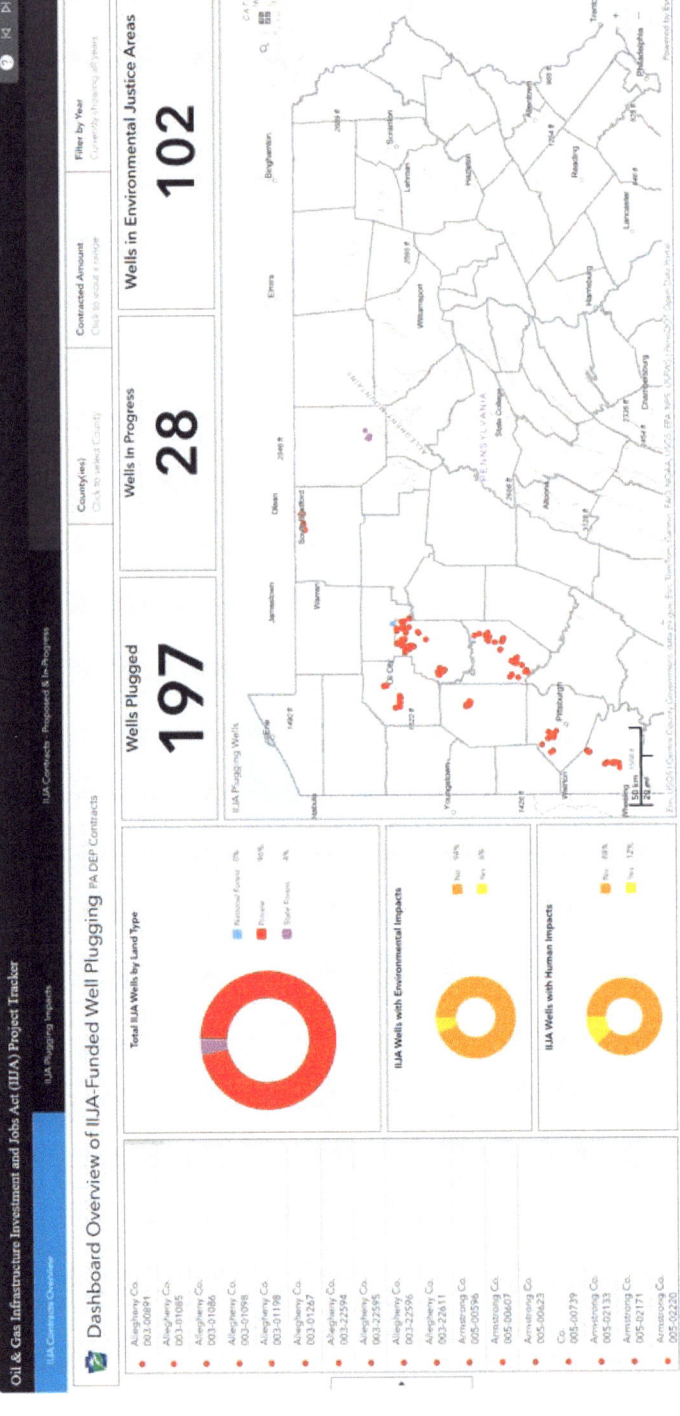

FIGURE 3-1 Snapshot of IIJA-funded well-plugging progress from Pennsylvania DEP's project tracker.
SOURCE: Pennsylvania Department of Environmental Protection, 2024.

filled with cement," and the wellbore must be static prior to placing the cement plug (16 TAC §3.14(d)(9)). Additionally, "all cement plugs, except the top plug, shall have sufficient slurry volume to fill 100 feet of hole, plus 10% for each 1,000 feet of depth from the ground surface to the bottom of the plug" (16 TAC §3.14(d)(11)) to mitigate potential contamination or pressure issues.

Sorrells also discussed several challenging issues that emerge during well-plugging projects and potential solutions. First, owing to various geological characteristics, specialty cement and mud may be necessary and circulation issues may arise. Second, crossflow, which can be determined with an ultrasonic log or a hydrophone, might require squeezing. Third, pressure issues may necessitate the use of kill mud, cement retainers, or packers. Fourth, cement retainers could be necessary for leaks in casing, while casing mills can help repair collapsed casings, and alignment tools and casing patches can help repair parted casings. He added that leaks often appear not only in the casing but also in packers or tubing, and "push plugs" can be used to address leaks that occur below the required squeeze zone. Sharing a series of project photographs, he emphasized that no single approach to well-plugging exists, and unique issues materialize with each well.

EVALUATING WELLBORE INTEGRITY AND SUBSURFACE CONDITIONS IN ALASKA'S ORPHANED WELLS

Bryan McLellan, Alaska Oil and Gas Conservation Commission (AOGCC),[4] explained that Alaska's first known wells were drilled in 1901, and the AOGCC was founded in 1958. Thus, most of the ~50 confirmed or suspected orphaned wells on state and private lands pre-date the AOGCC. Although Alaska did not have a state-level orphaned well program prior to the IIJA, he noted that the Bureau of Land Management (BLM) Legacy Well Program (which does not use IIJA funds) has been plugging a few wells per year in the National Petroleum Reserve–Alaska over the past 5–10 years.

McLellan highlighted the AOGCC's objectives for plugging wells, including to plug wells according to AOGCC regulations;[5] ensure that hydrocarbons and freshwater are confined to their indigenous strata and prevented from migrating into other strata or to the surface; prevent contamination of fresh water; and prevent waste of hydrocarbon resources. He stressed that achieving these objectives helps to prevent any risks to public safety. Therefore, one of the AOGCC's primary goals in allocating IIJA funds is to plug known wells to prevent additional contamination, methane emissions, or waste of hydrocarbons; then, if funds remain, subsequent goals are to remove or remediate reserve pits and remove debris from the sites, and to remediate existing soil or water contamination.

To select which wells to plug first, McLellan indicated that the AOGCC has modeled its scoring and ranking system after the RRC's, with categories that align with

---

[4] For more information about the AOGCC, see https://www.commerce.alaska.gov/web/aogcc/.
[5] For more information about the AOGCC's regulations, see https://www.commerce.alaska.gov/web/aogcc/StatutesandRegulations.aspx.

Alaska's specific issues. Similar to many other states' systems, Alaska's prioritization is risk-based, with leaking wells always being addressed first and efficient use of funding taking precedence (i.e., plugging all wells in a particular area instead of just those at high risk). He explained that wells are ranked and scored based on wellbore condition, reservoir risk, well location with respect to sensitive areas, and logistical and landowner considerations.

McLellan remarked that the first steps in evaluating wellbore integrity and subsurface conditions are to search the AOGCC's database for records, search U.S. Geological Survey or BLM files for wells drilled before Alaska's statehood in 1959, search museum records and historical photographs, conduct a title chain search and converse with landowners, and perform site visits. Key questions to consider include whether a well was plugged adequately and how deep the plugs should be. To determine minimum plug depth, he suggested using the pore-pressure/frac gradient tool (see Figure 3-2). In brief, if one starts with the pressure at the depth of the hydrocarbon reservoir and follows a gas gradient (grey line, which represents the pressure inside the wellbore, assuming the wellbore is full of gas) up to the point where the gas gradient intersects the frac gradient curve (orange line, which represents the fracture pressure in the formation outside the wellbore), this point of intersection represents the shallowest safe plug depth. If the plug is set shallower than this depth and the casing corrodes and develops a leak, the pressure inside the wellbore can fracture the rock outside the casing allowing gas to bypass the plug and leak to shallower zones or to the surface via the fractures. Ideally, then the plug would be set below that point, where the pressure inside the well could not exceed the fracture pressure, he said. Even if a leak occurred in the casing below the plug, the gas would likely stay in place (assuming there is annular cement outside the casing) because the rock strength can withstand the pressure below the plug.

The next step before placing plugs, McLellan continued, is for geologists to review mud logs and petrophysical data. To determine subsurface risk and priority, they aim to pinpoint where the freshwater is, if significant hydrocarbons and sources of overpressure exist, where the confining zones are, and how the pore-pressure/frac gradient curves can help. He noted that this geological analysis is then compared with wellbore conditions to understand risk and plugging priority, by reviewing cement records and bond logs, considering the location of existing plugs, and reviewing the well site survey of methane measurements and active leaks or signs of contamination.

PERSPECTIVES FROM THE BUREAU OF LAND MANAGEMENT

Matthew Warren, BLM, conveyed that BLM manages more than 245 million surface acres and more than 700 million subsurface acres across the United States. It has a trust responsibility for managing oil and gas on Indian lands, except for those of the Osage Nation. BLM is also the permitting agency for all federally managed minerals onshore under all surface estates (including those that are privately owned). He noted that BLM's orphaned well-plugging program emerged in 1976 when it started managing the U.S. onshore oil and gas program, and BLM currently manages more than 95,000 unplugged wellbores, including 8,500 federal idled wells, and 70 unplugged federal orphaned wells.

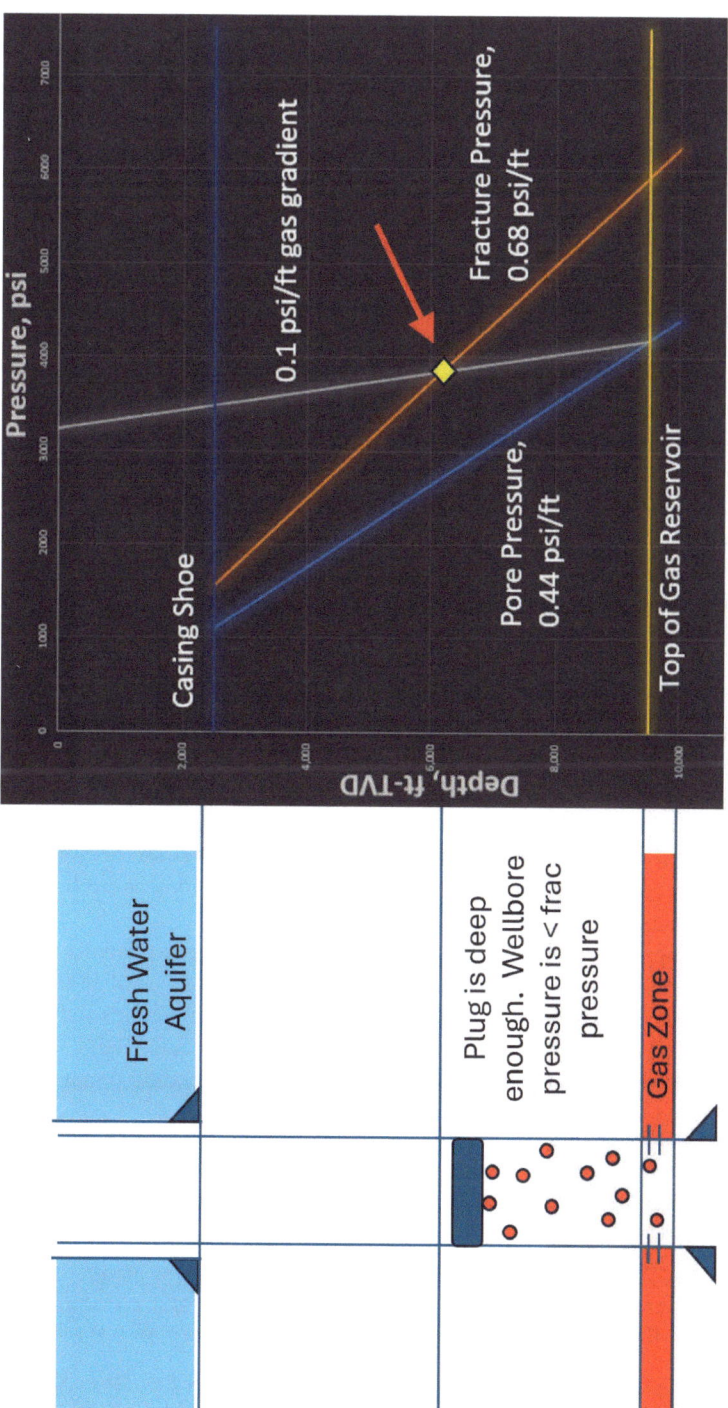

FIGURE 3-2 Using the pore-pressure/frac gradient to determine minimum plug depth.
SOURCE: McLellan, 2024.

Warren indicated that prioritization begins with consideration of the downhole integrity of idled wells, which the IIJA defines as those that have been inactive for 4 or more years and have no future beneficial use. BLM's 2020 Instruction Memorandum *Idled Well Reviews and Data Entry*[6] provides policy and guidance for conducting idled well reviews. He remarked that the highest priority generally is given to wells that have been idle for at least 50 years, followed by those that have been idle for 25–50 years. In the first half of fiscal year 2024, 427 idled wells were plugged or reclaimed; 252 idled wells were returned to production; and 609 new wells were added to the idled well count, resulting in a net decrease of 70 idled wells. He explained that BLM office staff are required to review 20% of idled wells each year via policy, and BLM field inspection staff review 40% of idled wells rated as highest priority each year.

Although BLM dedicates much time to idled wells, Warren said that BLM also plays a role in plugging orphaned wells. The IIJA defines orphaned wells as those with no "liable party"[7] responsible for plugging, which BLM must plug and reclaim. BLM also has agreements with Wyoming and New Mexico to plug wells, and BLM's 2021 Instruction Memorandum *Orphaned Well Identification, Prioritization, and Plugging and Reclamation*[8] provides policy and guidance for the identification and prioritization of orphaned wells. This document includes priority rankings—for example, whether the well is leaking at the surface or a problem with pressure exists, whether the wellbore configuration is known or unknown, how old the well is, whether surface contamination exists, if additional equipment or existing infrastructure are present, whether hydrogen sulfide is present, how close surface waters and water wells are, and whether a National Environmental Policy Act review has been completed.

BLM's specific regulations for plugging oil and gas wells can be found at 43 CFR 3172.12. Warren remarked that minimum requirements include that "all formations bearing usable-quality water, oil, gas, or geothermal resources, and/or a prospectively valuable deposit of minerals shall be protected"; that "cement is the default isolating medium"; and that, according to BLM processes, plugging operations must be approved by BLM prior to starting operations.

Before closing, Warren provided a brief overview of issues related to bonding. BLM's most recent leasing rule includes new bonding amounts: $500,000 for every statewide bond and $150,000 for every lease bond. BLM also maintains federal bonds and reviews Indian land bonds at the request of the Bureau of Indian Affairs. He pointed out that idled wells increase the amount of bond required, depending on an operator's percentage of idled wells.

---

[6] This Instruction Memorandum is available at https://www.blm.gov/policy/im-2020-006.

[7] For BLM, liable parties' responsibilities date back to when the well was drilled, even if the lease rights were sold or transferred, which differs from many states and may take a couple of years to pursue previous owners.

[8] This Instruction Memorandum is available at https://www.blm.gov/policy/im-2021-039#:~:text=This%20 IM%20defines%20an%20orphaned,cost%20for%20permanent%20well%20plugging%2C.

OPEN DISCUSSION

Radonjic moderated a discussion among the workshop speakers and participants. An online participant inquired about public partnerships for well-plugging activities in Pennsylvania. Hegburg replied that community outreach and engagement will be a key use of IIJA funds; his team has identified potential partners such as nongovernmental organizations and university extensions and will be conducting outreach sessions to discuss future plans, highlight the "plugging 101" process, and gather input.

James Saiers, Yale University, observed that some states prioritize plugging wells that exhibit a threat to groundwater or are in close proximity to water supply wells for households. He wondered if any states are conducting far-field monitoring of groundwater quality in areas with compromised wells. Kropatsch said that Wyoming is not currently conducting this kind of monitoring, primarily owing to the remoteness of its wells and their distance from water wells. Warren added that BLM is not conducting far-field monitoring either, but if indications are available from nearby wells, those become part of an analysis for prioritization. Hegburg explained that in Pennsylvania, wells are inspected as contracts are prepared. When a water supply is located nearby, pre-drill sampling is conducted to prepare for any issues that might arise during plugging. Eric van Oort, University of Texas at Austin, encouraged the use of data to make prioritization decisions and suggested incorporating artificial intelligence (AI) and machine learning to determine which wells are currently or could become problematic with leaks. Arthur mentioned that the Department of Energy is working on AI solutions for prioritization.

An online participant posed the following question: If a well operator is prevented from managing a well and Pennsylvania declares that well abandoned, does that operator have the right to rehabilitate the well instead of plugging it? Hegburg responded that stringent rules exist to declare a well "abandoned," and the landowner will be notified if a well deemed abandoned is going to be plugged. He urged operators to maintain familiarity with all state rules and regulations.

Dwayne Purvis, Purvis Energy Advisors, asked what data are helpful for labeling an orphaned well "documented" versus "undocumented" and whether some wells could be labeled "partially documented." Hegburg noted that he refers to the Department of the Interior's guidelines, which are based on locational data, to determine whether a well can be considered "documented." Workshop planning committee member Mary Kang, McGill University, added that data provided by the states on documented orphaned wells are often missing information on depth, type, and last production date. Purvis proposed developing a common taxonomy with clear thresholds by which wells can be counted as documented, undocumented, and partially documented. However, Radonjic highlighted the value of flexibility because states have different practices, cultural foci, and historical issues that affect what data can be collected.

4

Examples of Procedures and Best Practices for Wellbores

INTRODUCTION

During the third session of the workshop, moderated by workshop planning committee member Nathan Meehan, Texas A&M University, experts from industry shared examples of lessons learned and best practices for plugging orphaned wells.

CHALLENGES TO PLUGGING ORPHANED WELLS

Steve Plants, Plants and Goodwin, Inc., commented that while well-plugging may not be the most exciting part of the oil and gas industry, it is an important process for operators and regulators to avoid potentially costly mistakes. He also highlighted a historical disconnect between the industry and regulators and between the industry and environmental activists. However, he pointed out that the orphaned wells problem has improved cooperation and collaboration significantly, which could enable greater success.

Plants next discussed four challenges that contractors face when plugging orphaned wells. The first challenge relates to access. For example, as discussed in the previous session of the workshop, the surface owner is often not the same person or entity as the mineral rights owner, especially in the Appalachian Basin. Therefore, when orphaned wells are on private land, contractors' interactions with landowners are critical. Overhead power lines present another access issue when plugging wells, as utility companies may be hesitant to shut off power and relocate lines, especially given the associated high costs. Underground utilities (e.g., gas lines, water lines, sewer lines, fiber optic cable) also pose challenges. Furthermore, environmentally sensitive areas are important to consider when plugging wells, many of which were drilled long before people were concerned about preserving wetlands, for example. He added that topography (e.g., steep or remote areas) also creates access challenges, as do road stability (i.e., with equipment that weighs 100,000 lb per piece) and aboveground structures.

A second challenge that plugging contractors confront relates to well control, which Plants suggested is not discussed enough. Some of the elements of well control include casing integrity, a functioning wellhead, use of hot taps to access wells with unknown pressure, creation of barrier policies that protect contractors, and use of kill fluid.

Plants highlighted wellbore obstructions as a third challenge for contractors. In other words, not having well records; dealing with older wells; working with an open well casing that allows access to passerby who throw things in the well; finding unknown equipment left in the well at abandonment; encountering circulation issues; and lacking fishing, milling, and other specialty tools can all create difficult well-plugging situations.

Finally, cost is likely a significant challenge for contractors, and Plants explained that access issues, well control issues, and wellbore obstructions all increase cost. He suggested that cost per foot is an ineffective unit of measurement to determine the cost of plugging a well; for example, a shallow well might have more problems and thus cost more to plug than a deeper well. He proposed that a more effective strategy to determine cost starts with considering the goal—whether that means plugging wells at low cost, plugging wells only once, keeping workers safe, and/or meeting minimum standards (i.e., developing erosion and sediment controls, testing cement, tagging plugs, tracking waste disposal, and conditioning wellbores).

Plants suggested the following example operating procedure for plugging and abandonment: (1) verifying that all required permits are in place; (2) making notifications; (3) developing an emergency response plan; (4) conducting a pre-construction site visit; (5) completing access road and well pad construction; (6) assessing wellhead/cellar; (7) move the plugging rig and ancillary equipment on to the site and set up for operations; (8) cleaning out to total depth; (9) cementing bottomhole plug; (10) conducting wireline logs, cut casing, and/or shoot perforations based on well construction; (11) pulling uncemented casing; (12) pumping remaining cement plugs; and (13) reclaiming the location and access road.

EXAMPLE LEARNINGS FROM GULF OF MEXICO PLUGGING AND ABANDONMENT EXPERIENCE

Drew Hunger, Seashore Petroleum, LLC, reaffirmed Plants's comments on the challenges confronted by plugging contractors. For example, between 2011 and 2013, upon entering 10 wells plugged and abandoned by a previous operator that were leaking, he found at least 1,000 psi in four of the wells. In particular, the Gulf of Mexico has many temporarily abandoned wells with sustained casing pressure. He described two catastrophic failures related to carbon dioxide floods from previously plugged and abandoned wells in Louisiana and Mississippi, multiple plugging and abandonment failures in West Texas, and a significant well blowout in Colorado in 2023. He shared his concern that the number of plugged and abandoned wells now leaking across the United States is indicative of a significant problem.

Hunger then highlighted the importance of "bubbles." He explained that bubbles migrate upward due to buoyancy, regardless of pressure and fluid in a plugged and abandoned well. As bubbles migrate upward from a leaking lower plug and collect under an effective shallow plug, problems and risks arise. The bubbles build pressure equivalent to the reservoir pressure over time (in weeks or in decades) as the trapped gas volume increases and pushes fluid out of the well. Well control kick calculations indicated that if a stray zone has a 10 lb per gallon environment at 10,000 ft, 4,435 psi and 160,000 lb of force could build under a 7 5/8-inch plug if the fluids have been mostly displaced by gas. Therefore, he stressed that a well that leaks all the way to the surface is not nearly as concerning as one that leaks from the bottom to the top but has no surface expression. He also described sources for a high-pressure leak, including zones recharged from water drive, stray hydrocarbon zones at initial pressure, or future carbon dioxide injection zones. He added that the Gulf of Mexico does not have a frac gradient at shallow depths, which can increase the risk of failure.

With these issues in mind, Hunger pointed to the following suggestions for effective plugging and abandonment:

- Keeping hydrocarbons out of the wellbore from the start to prevent bubbles from forming in the well with this best accomplished by cementing productive zones;
- Designing for a long-term hydraulic seal, not just a mechanical plug;
- Preparing the pipe with surfactant to ensure a cement bond;
- Recognizing the benefits of cement additives and cement testing;
- Leaving the well alone until the cement reaches compression strength;
- Considering the impact of cool temperatures on cement and how long it takes to set;
- Checking the top of the cement after every plug;
- Never assuming that steel will maintain its pressure integrity over time;
- Never assuming that a mechanical barrier that relies on elastomers for a seal is a permanent barrier, as elastomers will fail eventually;
- Conducting a "bubble test" after every cement plug;
- Replacing plugs where they fail during the plugging process;
- Solving annular bubble problems deep in the well; and
- Avoiding the use of mud because it breaks down over time, loses its hydrostatic properties, and hinders the achievement of a hydraulic seal.[1]

Furthermore, to prevent plugged wells from leaking, Hunger suggested that well-plugging and abandonment be a recognized specialty for which people are well trained and that people designing plugging and abandonment plans as well as those pumping the cement be trained in cement and cement placement and understand the importance

---

[1] Hunger later offered clarification that mud is not used often in the Gulf of Mexico owing to reliance on a rigless plugging and abandonment approach and a dual barrier approach, with two cement plugs being placed before any open work is done with the well.

of going beyond the regulations with best practices. Hunger reported that most plugs in the Gulf of Mexico use Class H "neat" cement (i.e., no additives) and that more design should go into the cement. Echoing Plants, he noted that contract strategies that create an urgency to "rush to save" are dangerous (e.g., not waiting for the cement to reach compression strength). He also suggested that increased operator priority and oversight as well as strong regulations and inspections could be beneficial. By following this guidance, he suggested that high-cost, high-risk interventions could be avoided, thereby protecting the oil and gas industry's reputation as well as protecting key regions such as the Gulf of Mexico and inland waterways.

THE PROCESS: EVALUATION, PLANNING, AND EXECUTION

James Bolander, JLB Engineering, LLC, offered guidance from an operator's perspective but pointed out that in the case of orphaned wells, the operator's role and the regulator's role overlap. He highlighted three key stages of plugging orphaned wells: (1) evaluation, after identifying the well and the wellbore construction; (2) plan development, with particular attention to understanding the state regulatory structure for plugging and abandonment operations and identifying technical support and best practices; and (3) project execution, followed by inspection and post-completion monitoring.

Exploring these stages in more detail, Bolander first discussed the critical components of well identification, including age, drilling method, and purpose; well site location and condition; surface equipment, including the wellhead; and wellbore and subsurface conditions, including type, fluids, depth, and pressure. Potentially useful sources of this information include well histories, logs, and regulatory filings such as drilling reports, plugging permits, and plugging reports. When data are not available, he said that offset well evaluations and visual inspections of the well site could be useful. To evaluate well construction, he suggested gathering information on the casing use (e.g., conductor, surface, intermediate, or production) and casing depth (i.e., whether it meets state regulations), the perforated interval(s) and what can be isolated, the cement (i.e., whether the top of the cement is above areas that require isolation such as potential flow zones, usable water, producing zones, and corrosive zones), and other equipment in the well (e.g., tubing, packers, or casing equipment).

If the appropriate information has not been acquired during the well identification and wellbore construction evaluation to prepare a plugging plan, Bolander indicated that one could identify gaps and next steps, such as by conducting further inspections of the well and performing active testing. He suggested consideration of whether remediation is necessary prior to plugging the well (e.g., fish in the wellbore, casing repair). During the planning stage, he noted that states review the regulatory structure and determine zones that require isolation; review cement quality standards by type and compressive strength; determine placement, including plug type and use of the squeeze method; determine the thickness of the plugs; and consider static wellbore requirements to maintain well control. States also consider technical best practices from industry. For example, the American Petroleum Institute's (API's) Recommended Practice 65-3 discusses formations that require isolation as well as material considerations for barriers (e.g., environmental issues like the effect of temperature on cement). Industry Recommended Practice 27 identifies steps and best practices to perform permanent well abandonment, including a helpful pro/con list of barrier types. Lastly, the Ground Water Protection Council's *Well Integrity Regulatory Elements for Consideration*[2] discusses plugging operations, inspection elements, flexibility to use alternative methods and materials, and reporting requirements.

Reflecting on best practices for cement, Bolander suggested that the cement used for plugging and abandonment would ideally be of the same quality as the cement used for the primary cement job—that is, it would conform to API specifications, tailor to local conditions, meet compressive strength requirements, and meet free fluid content standards to minimize the potential for channeling. Meeting standards for mix water quality is also critical, he continued. To achieve operational success, he proposed following procedures for conditioning of the wellbore and requirements for maintaining static wellbore conditions prior to plug placement, including tagging or testing requirements for specific plugs, and tailoring plug placement based on varying well configurations. He suggested that flexibility in planning is key.

Finally, Bolander reiterated that inspection and monitoring play an important role in the project execution stage. Operators can consider when to be on site during the plugging process (e.g., for setting, tagging, and/or testing plugs), when corrective action is warranted, and what type of monitoring is appropriate (e.g., long-term monitoring in certain geographic regions and in active operating areas).

OPEN DISCUSSION

Meehan moderated a discussion among the workshop speakers and participants. He asked Hunger to elaborate on the relationship between hydraulic seals and mechanical plugs. Hunger explained that a mechanical plug will prevent any movement caused by shearing forces acting on the cross-sectional area of the plug along the interface between the cement plug and the pipe. A hydraulic seal will then prevent transmission of pressure or flow through the cement or at the interfaces of the cement plug and the pipe.

Bryan McLellan, Alaska Oil and Gas Conservation Commission, posed a question about how to reenter a well where the pressure below a shallow plug is unknown. Hunger said that after a dangerous experience, he learned never to enter an already-plugged well without controlling the pipe and the pressure and being prepared for a volume of gas to emerge. Thus, he began using a large-diameter coil tubing unit and, most importantly, started pilot milling through the cast iron bridge plugs to check for pressure. He pointed to these two actions as allowing for control of the pipe and control of the gas (at a lower volume).

Nick Gianoutsos, U.S. Geological Survey, asked how long well plugs could last, based on current standards. Thomas Lopez, ExxonMobil, explained that if the caprock is restored, a plug could last many thousands of years. Hunger suggested that plugs could

---

[2] To read this guidance, see https://www.gwpc.org/wp-content/uploads/2021/03/Well_Integrity_Elements_Revised_1_19_2021_002.pdf.

last until the geology heals itself (i.e., until the shales encroach and collapse the well); however, the cement may not last that long. Meehan elaborated that chemical processes that begin over the long-term help with solidification; however, unknown future issues (e.g., fault migration, carbon dioxide injection) could be problematic.

Workshop planning committee member Mary Kang, McGill University, suggested that if states recognize that some plugs will fail within 50 years, they could then develop a risk ranking system for long-term monitoring. Lopez expressed his support for this idea but emphasized how challenging it would be to predict which plugs will fail in the future—tens of thousands of wells are plugged in a lifetime, and finding leaking wells is very difficult. Michael Hickey, Colorado Energy and Carbon Management Commission, explained that, in some cases, plug failure could actually be a failure to *place* plugs. Danny Sorrells, Railroad Commission of Texas, noted that a "Railroad Commission plugger" is always on location from the start to the finish of a project in Texas to weigh plugs and monitor and document the process.

An online participant wondered how seasonality affects the plugging process. Plants described this as a critical issue in the Northeast because many of the roads where wells are located are secondary rural roads not built for heavy loads. Such roads are particularly vulnerable to damage during the "freeze/thaw" period, which might last from February through May, so projects located off of state highways that are less vulnerable to damage are conducted during that timeframe instead. He emphasized the value of well-developed plugging plans that consider seasonality to avoid unnecessary costs.

Karl Haase, U.S. Geological Survey, wondered if states or regulatory agencies are cataloguing the designs and technologies used for plugging wells. Hunger remarked that operators in the Gulf of Mexico have to submit wellbore sketches with their proposed plugging and abandonment procedures (and updates any time a change is made) as well as a weekly operations report and an end-of-operations report. Jesse Frederick, WZI, commented that California's WellSTAR system requires before-and-after diagrams of formations and cement calculations. Bolander noted that all states have these types of requirements, but Greg Lackey, National Energy Technology Laboratory, elaborated that although the records exist, they are not always in databases, and large-scale queries often require a review of paper records.

Adam Peltz, Environmental Defense Fund, inquired about training programs for members of the plugging industry. Bolander replied that Texas's regulations require that plugging contractors have specific qualifications. Plants said that the governor of Pennsylvania has set aside money for formal training; furthermore, Plants and Goodwin, Inc., has a rigorous program that includes 6 months of short-service training and mentorship, after which competency has to be demonstrated before work in the field can begin. Hunger described Superior Energy Services' hands-on training program in the Gulf of Mexico as an exceptional model. It had a test well, temporary living quarters, and all the necessary plugging and abandonment equipment in a fenced-in area, where new employees worked 14 days on, 7 days off to train for future offshore work. He suggested that such a model also helps prevent continuous turnover in the industry. Meehan observed that drilling courses at Texas A&M do not dedicate much time to decommissioning, and opportunity exists to expand educational offerings in general.

5

Environmental Risks and Monitoring

INTRODUCTION

The fourth session of the workshop, moderated by workshop planning committee member Mary Kang, McGill University, discussed both the environmental risks posed by leaking wells and opportunities for monitoring to mitigate related issues. She indicated that although the actual counts remain highly uncertain as wells continue to be documented, ~400,000 "nonproducing" oil and gas wells are located in Canada, and ~4 million nonproducing oil and gas wells are located in the United States. Orphaned wells comprise a subset of these nonproducing wells. Given the high number of wells, she said that managing and monitoring associated environmental risks is critical. When wells remain unplugged, fluids can migrate through pathways and affect both groundwater and air quality (see Figure 5-1), while risks of explosion also present safety concerns.

Kang noted that emissions of methane are of particular concern. Boutot and colleagues (2022) found that 123,318 documented orphaned wells had methane emissions that comprised 5–6% of total methane emissions from all abandoned wells in the United States. She explained, however, that uncertainties about methane emissions persist because the total number of wells remains unknown, and unmeasured or undetected high-emitting nonproducing wells likely exist. Many well attributes could contribute to the amount of methane released; for instance, some studies have captured data that suggest that well depth and age influence methane emissions. Thus far, however, she said that the only consistent contributing factor across methane studies is geographical area.

Kang remarked that isolating individual contributing factors is difficult because methane emission rates are driven by multiple processes. For example, when measuring methane emissions from the surface casing vent and from the wellhead separately, different trends are observed. She described a surface casing vent flow and gas migration dataset in Alberta that provides information about well integrity, but upon conducting

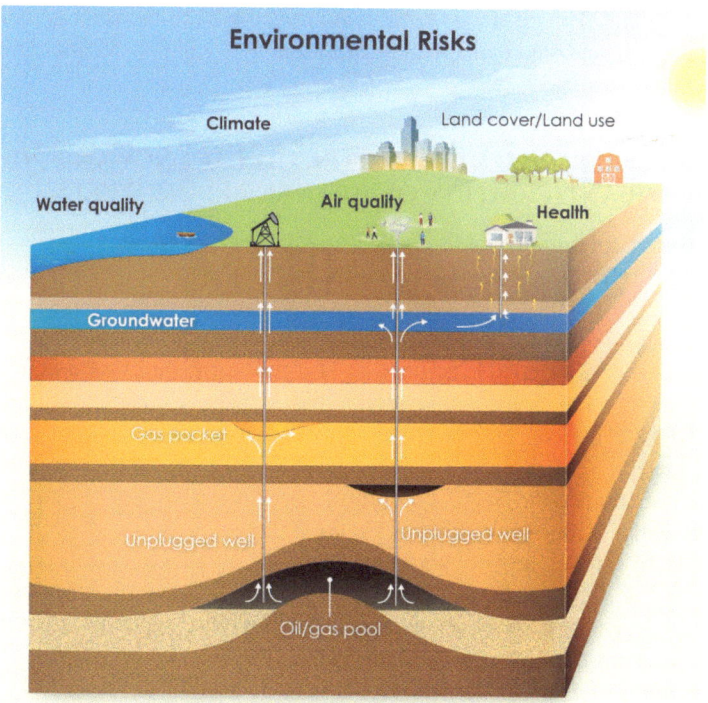

FIGURE 5-1 Environmental risks associated with unplugged wells.
SOURCE: Kang, 2024 (modified from Kang et al., 2023).

measurements, 68% of what her team found was not included in this dataset (Bowman et al., 2023). She stressed, however, that this difference likely is due to temporal variability.

Moving from a discussion of methane emissions to groundwater contamination, Kang noted that groundwater typically moves slowly, and different types of contaminants exist (e.g., some dissolve in water, but others do not). Cleaning up contaminated groundwater is both difficult and expensive.[1] She explained that the average groundwater well depth in the United States is ~230 feet, while oil and gas wells might have a depth of a few thousand feet (Perrone and Jasechko, 2017). However, groundwater wells are now being drilled deeper, and what constitutes "protected waters" might need to be redefined.

Before the first presentation of the session began, Kang invited speakers to consider the following questions: How confident are we with national estimates of orphaned and abandoned well methane emissions? Can methane emissions be used as a proxy for other environmental impacts? Can we use well attributes to identify potentially

---

[1] For more information on groundwater, see https://www.epa.gov/superfund/superfund-groundwater-introduction#:~:text=When%20contaminated%20oil%20or%20chemical,contaminated%20by%20the%20polluted%20groundwater.

high-emitting wells? How can we identify and remediate groundwater contamination? Where are the data and data management gaps, and how can new technologies be used to address them?

THE ENVIRONMENTAL PROTECTION AGENCY AND THE QUANTIFICATION OF METHANE FROM WELLS

Sarah Busch, Environmental Protection Agency (EPA), provided an overview of the greenhouse gas (GHG) data that EPA is collecting on abandoned wells. She explained that EPA's Inventory of U.S. Greenhouse Gas Emissions and Sinks (GHGI)[2] and GHG Reporting Program (GHGRP)[3] are complementary programs, as the latter provides key input to the former. The GHGI includes a "holistic" estimate of emissions and sinks across all sectors of the U.S. economy, including a category on abandoned oil and gas wells; the GHGRP collects emissions data only from large GHG-emitting facilities in the United States.

Busch first outlined the GHGI's multipronged classification of "abandoned wells." Abandoned wells include (1) wells that are not plugged but have no recent production (i.e., idle, inactive, temporarily abandoned, dormant, shut-in), (2) wells with no recent production and no responsible operator (i.e., orphaned, abandoned, deserted, long-term idle), and (3) wells that have been plugged. The GHGI's estimate is based on ~3.9 million wells, ~3 million of which are oil wells; orphaned wells comprise a subset of these wells, but the GHGI does not include a specific estimate of orphaned well count. She noted that over the past 30 years, the overall count of abandoned oil and gas wells has increased significantly, which corresponds to an increase in carbon dioxide and methane emissions.

Busch explained that EPA developed methane emission factors at both the national level and specifically for the Appalachian region using data from Kang and colleagues (2016) and Townsend-Small and colleagues (2016). Well counts were calculated using Enverus data to determine the fraction of plugged to unplugged abandoned wells. EPA then developed state-level annual counts of abandoned wells for 1990–2022 by summing an annual estimate of abandoned wells in the Enverus dataset and an estimate of wells not in that dataset (i.e., from historical records from state agencies and the U.S. Geological Survey). She noted that EPA continues to assess new data and feedback on how to improve the abandoned oil and gas well emissions and activity data, and the GHGI will be updated as new information becomes available. For example, several new data sources are under development, including the Department of the Interior's (DOI's) orphaned well database, the National Energy Technology Laboratory's (NETL's) marginal conventional well database, and Subpart W[4] data improvements. She indicated that Subpart W focuses specifically on the wells from large-producing facilities—that

---

[2] For more information on the GHGI, see https://www.epa.gov/ghgemissions/inventory-us-greenhouse-gas-emissions-and-sinks.

[3] For more information on the GHGRP, see https://www.epa.gov/ghgreporting.

[4] Subpart W refers to the reporting requirements for wells for petroleum and natural gas systems (40 CFR Part 98, Subpart W), see https://www.epa.gov/ghgreporting/rulemaking-notices-ghg-reporting.

is, greater than 25,000 metric tons of carbon dioxide equivalent, or about 50% of the producing wells in the United States. Beginning in 2025, she added, EPA will begin collecting more well-level estimates of pre-plugging emissions.

Busch next commented on three EPA regulations, not all of which have been finalized (see Figure 5-2). In particular, EPA will collect information on wells in the onshore petroleum and natural gas production sector, in the offshore petroleum and natural gas production sector, and in the underground storage sector.

She conveyed that information on well emissions from onshore production can be collected in several ways, such as via measurement and/or engineering calculations from equipment leaks, well venting for liquids unloading, and completions and workovers with and without hydraulic fracturing. To gather insight into well emissions from offshore production, data and quantification methods from the Bureau of Ocean Energy Management's Outer Continental Shelf Emissions Inventory are used. This year, Subpart W will collect information on the quantity of natural gas and crude oil and condensate sent to sale for each permanently shut-in or plugged well. For underground storage wells, well-level information is not available, but facility-level equipment leak emissions information will be categorized (e.g., by storage wellhead).

Busch explained that Subpart W was updated in response to new authorities under the Clean Air Act (Section 136) to reduce methane emissions from oil and gas via the Methane Emissions Reduction Program. These new authorities included creating an incentive program for financial and technical assistance. Accordingly, NETL developed the Methane Measurement Guidelines for Marginal Conventional Wells.[5] Marginal wells are defined as those that produce less than or equal to 15 barrels of oil equivalent per day over the prior 12-month period; the minimum detection limit is less than 100 grams/hour. To quantify pre-plugging emissions, a sample could be taken directly from the well using high-flow sampling with flux chambers or bag sampling, or field measurements using drones or ground-based vehicles described in EPA OTM 33A. She added that remote sensing could be used in the future but only if it could achieve the minimum detection limit. Post-measurement requirements are detection-based only, she suggested the use of EPA Method 21 or OGI in this case. She noted that this is the minimum data reporting requirements; however, other data reporting is encouraged.

Busch pointed out that these new authorities under the Clean Air Act also included establishing the Waste Emissions Charge,[6] which was proposed in January 2024 but from which some wells would be exempt. EPA proposed to exempt emissions from wells that are permanently shut-in or plugged in accordance with all applicable closure requirements, and the exempted emissions are limited to wellhead emissions (i.e., leaks, liquids unloading, and workovers).

---

[5] The guidelines are available at https://netl.doe.gov/methane-emissions-reduction-program.

[6] For more information about the Waste Emissions Charge, see https://www.epa.gov/inflation-reduction-act/waste-emissions-charge.

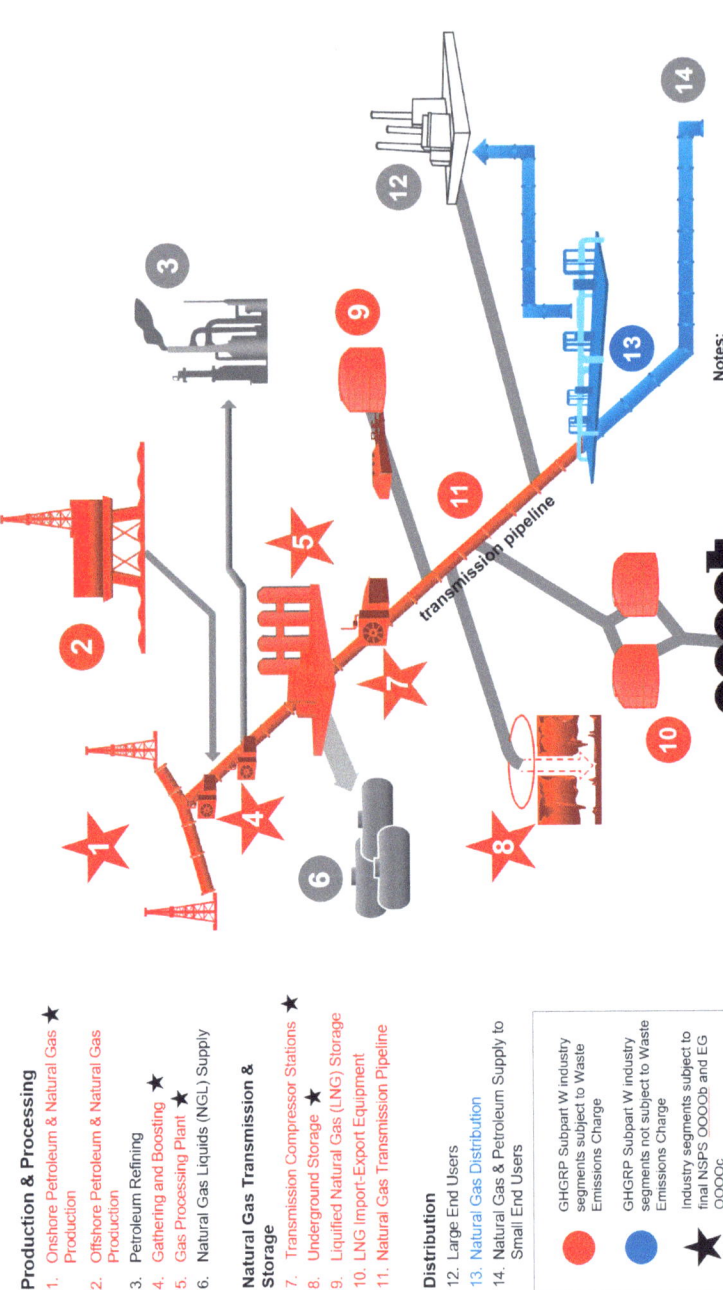

FIGURE 5-2 GHGRP Subpart W industry segments subject to Waste Emissions Charge (red), GHGRP Subpart W industry segments not subject to Waste Emissions Charge (blue), and industry segments subject to new source performance standards and emissions guidelines (black star). SOURCE: Environmental Protection Agency, 2024.

POTENTIAL METHODS FOR QUANTIFYING METHANE EMISSIONS FROM ABANDONED OIL AND GAS WELLS

James France, Environmental Defense Fund, explained that when trying to measure methane, the environments of the measurement regions might vary significantly. In some cases, the location is unknown. However, when the location is known, questions arise about whether single wells, multiple wells, or large clusters of wells need to be quantified; if getting close to the infrastructure is feasible; if one has a clear line of sight; and whether changeable weather conditions could create issues.

France indicated that for isolated infrastructure in remote locations with difficult and variable terrain, bespoke chamber methods are useful to measure methane emissions. However, he said that different environments dictate different measurement methodologies. For example, Azerbaijan has semi-active infrastructure and redundant infrastructure together, so a very different measurement approach would be used. He suggested choosing a quantification methodology based on the scale of the environment. For instance, component-level methods to detect emission (e.g., high-flow sample bagging and an optical gas imaging camera) are appropriate to quantify known leaks that can be visualized or measured directly. At the facility level, where a cluster of infrastructure might exist, downwind methodologies (e.g., open-path measurements) might be useful to capture a full plume or a subsection of a plume. He noted that these types of methodologies are useful in generating a total quantification measurement and for monitoring long-term emissions trends but will not pinpoint exact locations for mitigation. At the basin level, methodologies include the use of satellites and aircraft; however, information about the sources of the emissions is limited with these methodologies. He proposed that principles from facility-level quantification could be applied at the component level. Providing an example of using a Lagrangian dispersion model to help understand the expected emissions from an isolated abandoned well infrastructure, he suggested that open-path methods rather than direct measurements could be viable solutions with the right technologies.

France next provided a brief overview of five specific methodologies that could be used to measure methane emissions from abandoned oil and gas infrastructure in the future, starting with the most precise and the most expensive option. First, he described bespoke chamber systems as the "gold standard" for abandoned well quantification. Using these systems, one can isolate the emissions source completely, and the method is usually reliable and not influenced by other local sources. However, he noted that this method is time-consuming and expensive and often requires custom fabrication. Simpler off-the-shelf options (e.g., a high-flow sampler) could be appropriate in some situations, but then many emissions might not be detectable. Second, he then noted that in a few years, a small-scale eddy covariance system may provide quantification over several sites a day as a single-person set-up. However, he cautioned that these systems are difficult to install because they require consideration of the terrain and the footprint of the source, are scientifically challenging to use, have expensive instrument costs, and require a separate screening method. Third, he noted that EPA's Other Test Method 33A, which is being used to quantify downwind plumes from industrial sources, could be adapted for abandoned oil and gas infrastructure with modern, highly portable parts per billion–level precision instrumentation. This instrumentation is also expensive but could collect sufficient data in 30 minutes using an open-path methodology. Emission rates would still have medium-to-large uncertainties, he continued, but a deluxe edition could release tracer gas at the same point as the emission source to eliminate any questions about the influence of the wind. Fourth, he explained that quantified optical gas imaging systems are straightforward to use and allow for screening, localization, and quantification in a single visit with a single instrument. The systems do not offer robust quantification, though, owing in part to interference issues. Fifth, he indicated that open-path tunable diode laser absorption spectroscopy (TDLAS) currently is the only remote sensing option for abandoned wells. Drone-mounted TDLAS could screen for the largest emitters, especially when combined with magnetometry surveys. Limited literature is available on their effectiveness; although these systems are not very precise, he mentioned that their best potential is for large-scale emission detection and for methane mitigation planning.

Before concluding his presentation, France presented a brief overview of available sensors. Laser cavity systems, which are not yet widespread in industry, could be useful if open-path measurements are used more often. These systems are very precise (with an accuracy of 1 ppb in atmospheric methane) but very expensive (at least $20,000–$30,000). Less expensive, smaller options that can be mounted on drones are also emerging. He added that commercial metal oxide sensors could be useful for long-term monitoring if attached directly to infrastructure. He reiterated that although the bespoke chamber system remains the gold standard for quantification, for cost-effective "bucketing" of emissions for mitigation planning and calculating generalized emission distributions, open-path options are viable and hopefully will continue to improve as technology and platforms progress.

METHANE EMISSIONS AND WATER QUALITY IMPACTS AROUND ORPHANED AND ABANDONED HYDROCARBON WELLS

Susan Brantley, Pennsylvania State University, noted that DOI is asking each recipient of an Infrastructure Investment and Jobs Act (IIJA) grant to (1) document the methodologies that the state will use to measure and track groundwater and surface water contamination related to orphaned wells, including how it will assess the effectiveness of plugging to address this contamination; and (2) document both pre- and post-plugging values of gaseous emissions and water contamination.

To assess the effectiveness of plugging, Brantley suggested first considering at what point water can best be sampled and analyzed to measure and track contamination, and by whom. Proving that plugging a well fixed a particular problem is difficult, she explained, owing to a variety in sources of contamination and seasonal changes. Therefore, several lines of evidence—water chemistry, gas chemistry and isotopes, pump tests, three-dimensional geologic mapping, geochemical modeling, and hydrological modeling—would better attribute causation. She indicated that the "gold standard" is to collect timeline data both before and after plugging a well, with professional hydrogeologists and certified laboratories conducting all sampling and analysis activities.

Brantley stated that the second step in assessing the effectiveness of plugging is to determine what could be chemically analyzed in the water—for example, gases, salt cations and anions, and/or metals and toxic species. In other words, one would determine which primary contaminants are released by modern or legacy oil and gas wells. She mentioned that as of 2023, 50% of the Pennsylvania Water Supply Determination Letters cited methane release, indicating that this is a problem caused by shale gas well development. However, many sources of methane are unrelated to oil and gas—for example, biogenic gas; landfills; or thermogenic gas, which often emits naturally from fractured shale. She explained that documenting that the methane is actually from a leaking well often requires analyzing isotopes and ethane, which is a very expensive and non-definitive process. Another key issue to consider when determining what could be chemically analyzed in the water relates to the secondary contaminants that are emitted from the subsurface rock when the primary contaminants are released. For example, when methane leaks from a borehole, secondary contaminants can be discharged. The methane can be oxidized and coupled by bacteria to reduction of sulfate; when sulfate is reduced, it becomes hydrogen sulfide—a toxic gas. Methane also can be oxidized to carbon dioxide at the same time that iron oxides are reduced to iron(II); therefore, she continued, many iron oxides are present in the subsurface and often contain other metals (e.g., arsenic) that can be released into solution at the same time and be considered secondary contaminants.

Brantley noted that after methane, the second most frequently identified contaminant group from shale gas in the Pennsylvania Water Supply Determination Letters includes iron, manganese, and turbidity. She observed that distinguishing contaminants from natural iron and manganese is difficult, reinforcing the value of collecting data both before and after plugging to make a determination. The third, most commonly identified contaminant group includes salt species in oilfield brines, because large volumes of salty water come up with the oil and gas. Thus, she explained that seeing high total-dissolved solids, chloride, and barium is common in shale gas regions. However, as in the previous case, other sources exist such as road salts, naturally upwelling brine, and non-oil and non-gas pollution; to determine whether brine has impacted a water sample, one can use ratios of indicator elements such as chloride or bromide (Woda et al., 2018). She added that considering what is in these brines is worthwhile. Brines in different basins across the United States differ; for example, the Marcellus in Pennsylvania has some of the saltiest water. In addition to sodium chloride and indicator elements, it could include organics, toxic elements, and radium (Barbot et al., 2013; Rowan et al., 2011).

Because some toxic elements might be elevated slightly in concentration, Brantley continued, another important issue to consider when determining what could be chemically analyzed in the water relates to identifying toxic species for local residents. Using a back-of-the-envelope calculation to determine the chloride concentration in drinking water if the water were contaminated by this "average Pennsylvania brine" and assuming that each metal reached EPA's maximum contamination limit, Shaheen and colleagues (2022) found that some of the toxic elements reached concentrations at which the person drinking the water would not taste the salt.

Brantley remarked that in the third step of assessing the effectiveness of plugging, one chooses where water would be sampled. She referenced a study on the potential use of surface water to detect methane contamination from oil and gas wells (Wendt et al., 2018; see Figure 5-3). Samples were collected at 131 sites across the northern

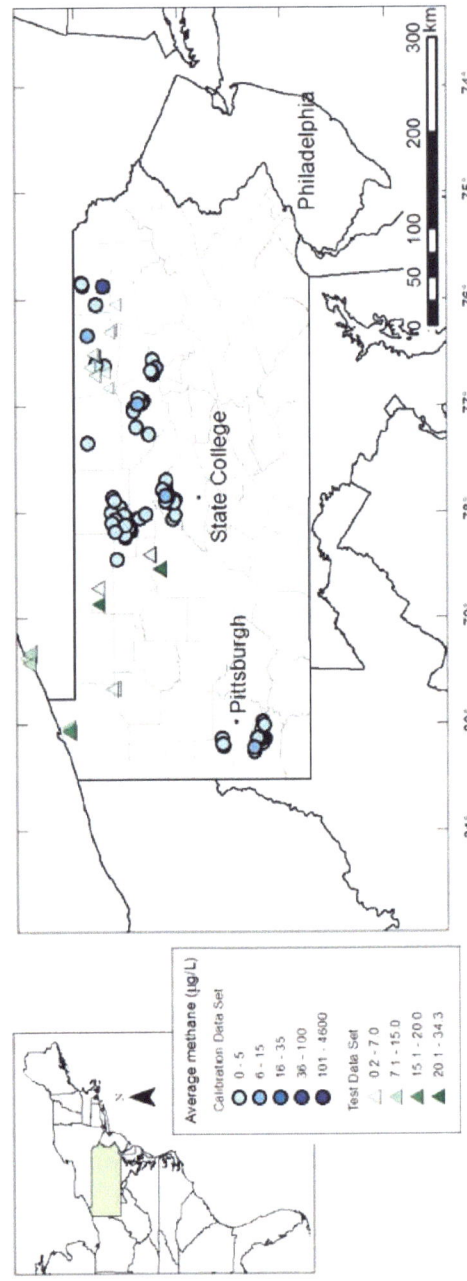

FIGURE 5-3 Locations and values of methane measured at sites in reconnaissance (circles) and contamination-targeted (triangles) datasets.
SOURCE: Wendt et al., 2018.

Appalachian Basin; 470 of 479 surface samples were supersaturated with respect to methane in the atmosphere. The study suggested that for non-wetland sites, concentrations above 4 g CH_4/L could be from a leaking gas well, shallow organic-rich shale, coal releases, or a landfill. The study found that 12 of 41 non-wetland sites near active, plugged, orphaned, or abandoned oil and gas wells were above this threshold of 4 g CH_4/L. The study thus points to surface waters as evidence of methane migration from leaking wells; however, she suggested that wetlands and other types of sites like landfills be avoided, and dilution usually hides contamination.

Moving from a discussion of sampling surface water to sampling groundwater, Brantley explained that in some terrains, contaminants are not observed in groundwater close to the source; instead, they are observed 1–2 km in the distance owing to subsurface fracturing, faults, and folding. She added that geology is complicated in terms of finding connections between a source and contamination. In another example, she indicated that homeowner wells located 1–3 km from shale gas wells were found to be contaminated with methane and an aqueous pollutant—one of the farthest documented distances of migration of contamination associated with a shale gas well (Llewellyn et al., 2015). Therefore, she suggested sampling water from drinking water wells and surface waters within 1–3 km, preferably downgradient and at the same or lower elevations as a legacy well.

LEVERAGING PUBLICLY AVAILABLE DATA TO UNDERSTAND WELL-INTEGRITY RISKS

Greg Lackey, NETL, provided an overview of a multiyear effort to analyze large datasets to understand well-integrity risks. He first explained that when aggregating state plugging prioritization systems, he and his colleagues found that a well's distance to a sensitive receptor is the highest priority, followed by a well's leaking status. Although a leak can be caused by a well-integrity issue, especially in older undocumented wells, he pointed out that integrity issues do not always result in an emission of gas. He noted that with nearly 150,000 documented orphaned wells in the United States, conducting inspections and collecting measurements from each well to inform prioritization efforts is not currently possible.

However, Lackey commented that data from well-integrity monitoring programs provide insight into leakage. He said that leaks in wells begin with a source and a leakage pathway, which varies based on well construction. The pathway could be to the surface, to groundwater, or to the atmosphere. For example, the source of a leak could be an over-pressured zone, and the leakage pathway to the surface could be from an uncemented portion of the annulus, a micro-annulus, gas channels, fractures, or a casing leak. Pathways to groundwater include a fault or preferential pathway along the wellbore into which gas migrating upward can escape. For example, in Colorado, some areas have had sustained casing pressure that has built up so much that it has displaced fluid from the surface casing. He emphasized that this evidence demonstrates the value of monitoring for sustained casing pressure specifically and of developing well-integrity monitoring programs more generally. In terms of pathways to the atmosphere, he noted that if annuli are left open, fluids that are transmitted into the wellbore will escape into the atmosphere. Other pathways to the atmosphere include faulty fittings on a wellhead that could transmit production gas.

Lackey next described results from a 2021 study that aggregated testing data from Pennsylvania, Colorado, and New Mexico. These data were analyzed based on American Petroleum Institute (API)'s protocol to interpret annular pressures and diagnose sustained casing pressure (Lackey et al., 2021). The study found that the frequency of integrity issues among wells varies widely across regions—for example, in Colorado, integrity issues occurred in 0.3–26.5% of the wells, depending on the basin. Since Colorado began requiring testing for all wells, the 26.5% has decreased to 17.1%. He explained that other case studies have highlighted the varying causes of integrity issues in different regions. For example, in an area of Alberta with a cluster of integrity issues, the majority of surface casing annuli were found to have mostly intermediate gas (Szatkowski et al., 2002). In a study of wells in Colorado, the gas isotopically matched the production gas, with the cement top above that zone, in ~70% of the ~2,000 wells analyzed (Lackey et al., 2022). He indicated that these wells likely had micro-annuli that were causing sustained casing pressure.

Lackey specified that for large datasets like those used in the previous case studies, machine learning models can help to make useful predictions about integrity issues and provide insight into the drivers of integrity issues. Classification models are particularly effective in screening for wells that may have sustained casing pressure or casing vent flow. However, these models cannot yet predict the magnitude of such issues, and they might provide conflicting information. Yet, he said that the fact that causes of leaks vary between studies is expected given the variation in well types, sources of leaks, and leakage pathways.

Lackey also mentioned that in these modern datasets, the relationship between well age and leakage is often reversed, in part because more integrity issues are observed in newer horizontal and directional wells and because the datasets are biased toward newer wells with owners. Furthermore, he noted that well-integrity issues tend to be spatially clustered, given that geology varies by location and wells are designed for specific geologies (Bachu, 2017; Sandl et al., 2021). Therefore, if the specific geology can amplify clustering and frequency of integrity issues, he said that well-integrity monitoring programs are especially helpful in identifying "hotspot" areas for sustained casing pressure. He added that because wells near one another are likely to be similar and because leaking wells could transmit fluids to nearby wells, modeling based on spatial clustering to predict leak potential could be valuable for prioritizing the inspection of orphaned wells, especially when minimal information is available on the wells themselves beyond their location (Lackey et al., 2024).

Lackey remarked that whether the lessons learned from these large datasets of newer wells apply to older orphaned wells remains to be seen. With DOI's Orphaned Wells Program, an opportunity exists to collect emissions measurements from orphaned wells and to determine whether they were leaking when they were abandoned. He highlighted the importance of aggregating these data (as well as data from wells without emissions) to better understand well integrity and well leakage. For example, in a recent study conducted by the U.S. Geological Survey, emissions measurements were aggregated to reveal that high emissions are being found in wells located in thermogenic gas reservoirs (Gianoutsos et al., 2024). The Department of Energy (DOE) also is contributing to this goal to leverage data by developing methods to better document and characterize orphaned wells (including by creating tools to digitize records and synthesize multiple forms of data) and by developing noninvasive well-characterization techniques.

Lackey underscored the value of continuing to collaborate with the states to synthesize well data (e.g., compliance inspection data) with other information. Furthermore, he pointed out that all of this information also could be useful for those in other industries working to understand legacy well leakage risks before using the subsurface for other purposes in the future.

INFORMING GROUNDWATER-QUALITY MONITORING WITH MODELS

James Saiers, Yale University, observed that although groundwater quality monitoring is critical to protecting aquifers and human health, it is expensive and rarely used to assess contamination from orphaned wells. For groundwater monitoring around orphaned wells to become practical, he said that measurements would target areas with the greatest vulnerability to contamination. Thus, he presented a vulnerability framework that could inform the design of groundwater monitoring plans to evaluate the impacts of orphaned wells.

Saiers first provided a definition of vulnerability: the probability of drinking water impairment at a receptor location (e.g., a household drinking water well) in the event of a contaminant release from a source (e.g., a leaking orphaned well). He explained that when contaminants are released from a source, they do not move randomly; rather, they move in a preferred direction according to physical flow and transport processes. Therefore, vulnerability depends on the spatial relationship between sources and receptors as well as groundwater flow patterns and rates.

Saiers indicated that vulnerability can be estimated with hydrologic models. Model-based vulnerability assessments can help address the following questions related to groundwater quality monitoring around orphaned wells: (1) Which domestic water wells are most likely to be affected by orphaned well contamination? (2) When will contamination from orphaned wells likely be detected at water supply levels? (3) Where can monitoring wells be installed to detect and characterize contamination from orphaned wells?

Saiers elaborated on this concept of the vulnerability framework by providing examples from studies conducted in the Appalachian Basin, which considered the vulnerability of residential drinking water supplies to wastewater spills. Although the source of contamination in these examples was an unconventional oil and gas well pad (Soriano et al., 2020), he pointed out that a similar analysis could be done with orphaned wells as the source of contamination. The first step in this watershed-level vulnerability assessment was to develop and calibrate a groundwater flow model to predict the distribution in hydraulic heads or groundwater levels. Groundwater flow patterns then could be deduced, and contaminant migration could be simulated. He indicated that this model was calibrated by identifying an ensemble of simulations that allowed flow model predictions to match groundwater discharges to stream and measured aquifer water levels. Two thousand realizations were run with this model to best assess the uncertainty in the model parameterization on groundwater flow (Soriano et al., 2020, 2022).

Once the groundwater flow model was created, Saiers noted that the next step was to couple that model's output with a particle tracking model. Virtual particles were injected at the well pad, and their trajectory was followed as they moved through the groundwater flow system. These particle tracks indicated where and how far the contaminant might travel based on groundwater flow. He said that this particle tracking simulation was conducted 2,000 times (for each realization of the groundwater flow model) and produced 2,000 hydrologically possible trajectories for contaminant migration away from the well pad (Soriano et al., 2020, 2022).

The following step, Saiers continued, was to use these particle tracks to estimate vulnerability. Vulnerability was defined, for each grid cell in the model domain, as the number of simulations in which a particle track intersected with the grid cell out of the total number of realizations. He explained that vulnerability ranges from 0 to 1; higher values indicate greater agreement among the realizations that a grid cell falls on an advective transport path from the well pad (Soriano et al., 2020, 2022). For a particular snapshot in time, the vulnerability predictions revealed which areas might be impacted if a contaminant was released from a well pad. He pointed out that the vulnerability plumes were not all oriented in the same way, which reflects the roles of topography, geology, and the stream river network in creating complex groundwater flow patterns.

Consequently, Saiers noted that the next phases of the vulnerability assessment were to scale up from the watershed level to all of Bradford County, Pennsylvania as well as to compare the vulnerability estimates to actual measurements of drinking water quality. Ninety-one samples of drinking water were collected; the use of elemental ratios helped to distinguish non-impacted groundwater from water that had been influenced by inputs of produced water. Many samples that appeared to be vulnerable to contamination were not actually impacted, which was expected given that spills on well pads are fairly rare. Furthermore, six of eight samples from the shale gas produced water region were from groundwater wells identified as being vulnerable, which suggested that a link might exist between spills at well sites and contamination. He and his team then looked for written records of spills near vulnerable domestic wells impaired by produced water signatures: five of the six vulnerable wells were near well pads where spills had been recorded.

With this information, Saiers and his team modified the vulnerability approach so that it could scale further. It was applied to a large region across Ohio, West Virginia, and Pennsylvania, where ~10,000 wells had been drilled into the Marcellus and Utica shale and where 1.5 million people rely on private water wells for their drinking water. Their analysis predicted that 4% of this region is vulnerable to contamination from unconventional oil and gas, which includes ~2% of the population, although the vulnerability is distributed unevenly (Soriano et al., 2022). He noted that these vulnerable areas thus could be monitored when concerns about contamination arise.

In closing, Saiers emphasized that monitoring cannot be conducted everywhere owing to high costs and limited accessibility. This vulnerability framework can be used to optimize locations for groundwater sampling and to characterize the extent of contamination in a cost-effective way; to identify domestic wells of greatest risk for targeted monitoring and preventative action; and to support contaminant-source attribution analyses. He underscored that vulnerability analyses can be conducted with free, public domain software (e.g., MODFLOW, MODPATH)[7] and by professionals employed in consulting firms and governmental scientific and regulatory agencies.

OPEN DISCUSSION

Kang moderated a discussion among the workshop speakers and participants. A participant observed the significant spread on the particle tracking in Saiers's models and asked if additional measurements and conditioning data would narrow the range of potentially affected areas. Saiers indicated that the spread reflects uncertainty, but having more data to characterize the hydraulic conductivity field could reduce this uncertainty. He added that because the plumes exclude a significant area, one knows where *not* to spend time monitoring.

An online participant posed a question about whether North American plugging and monitoring best practices are transferable to other extractive industries and to other continents. Kang replied that although they are transferable to some extent, additional studies could confirm *how* transferable. France added that much insight can transfer, especially in terms of the methodologies used for emissions monitoring; however, different countries have different regulatory environments and different practices to consider. Brantley elaborated that geology also varies by location. She provided an example based on a study of fugitive methane incidents related to shale gas development in the southwest and northeast areas of Pennsylvania. The methane incidents were higher in the northeast than in the southwest, where much of the intermediate-depth gas had already been extracted. Thus, in addition to understanding the geology of a particular area, she said that it is important to understand legacy extraction activities.

Dwayne Purvis, Purvis Energy Advisors, asked if different standards are needed for decommissioning to prevent leaks from shale wells that will repressurize over years to decades. Lackey responded that data from the long-term monitoring of plugged wells could address this question. Workshop planning committee member Nathan Meehan, Texas A&M University, pointed out that if significant active migration occurs in producing wells, the problem likely will persist in abandoned wells, and he suggested that an alternate approach for the repair and plugging may be needed. Lackey added that some states that conduct integrity monitoring require operators to repair annular problems. He agreed that knowing which wells have these problems is critically important and proposed that remediation efforts occur prior to abandonment, but that is not clear in all jurisdictions' requirements.

---

[7] More information on MODFLOW can be found at https://www.usgs.gov/mission-areas/water-resources/science/modflow-and-related-programs.

Kang invited state leaders to share their perspectives on methane monitoring and groundwater modeling approaches. Michael Hickey, Colorado Energy and Carbon Management Commission, emphasized that oil and gas systems involve much more than just a well. For example, gas could emerge in a field 1 mile from a well or a tank battery owing to a leak in a flow line. Another issue is that wells do not flow uniformly; they surge. Therefore, he said that how long to monitor wells remains a very difficult question. Lackey noted that he and his team are expanding their models to include the whole well pad, and Busch reflected on the potential for improvements in Subpart W as well as the value of capturing more emissions data.

Bryan McLellan, Alaska Oil and Gas Conservation Commission, expressed his interest in evaluating the effectiveness of Alaska's plugging program and wondered if a simple, low-cost method exists to screen wells that have been plugged and abandoned. Kang replied that DOE is developing useful tools, and Lackey highlighted the FAST (Dubey et al., 2024) method as a low-cost way to measure emissions. However, he said that such a method, which focuses on emissions sources from the wellhead or an open hole, might not be appropriate if the goal is to verify already-plugged wells. McLellan inquired specifically about the use of remote sensing techniques, and Kang explained that drones might detect methane emissions to the atmosphere but might not be useful to detect subsurface impacts. France added that TDLAS has great potential to screen for very potent leaks; however, a higher-precision instrument could detect lower volumes of methane. Brantley remarked that collecting and analyzing samples is very expensive. She suggested instead spending federal funds on plugging wells and then conducting simple screening for methane. Hickey encouraged the use of soil gas sampling, which is a fairly straightforward and effective methane detection method being used at the county level in Colorado. Kang then asked about the use of saturated and unsaturated coupled models. Saiers explained that those multiphase flow models are critical if one is interested in the dissolved and gaseous states of methane, but they are difficult to scale and require greater characterization than single-phase flow models. Multiphase flow models could guide one where to look for methane, he continued, but measurements would still be needed for verification.

Don Hegburg, Pennsylvania Department of Environmental Protection, commented that the vulnerability framework presented by Saiers could be used to assist with well-plugging prioritization. He cautioned that when water supplies are sampled, they likely will exceed some standards because water supplies inherently are not well maintained; drinking water wells are subject to surface contamination that is not caused by oil and gas. Before allocating a significant amount of funding toward sampling, he suggested first conducting a basic geochemistry analysis or checking for high-conductivity subsurface areas around a well site.

Danny Sorrells, Railroad Commission of Texas (RRC), remarked that 8,674 orphaned wells are found across 10 districts in Texas. Each district has at least two forward-looking infrared cameras that check those wells regularly for leaks. Furthermore, all 180 of the RRC's oil and gas inspectors carry salinity meters and can check nearby creeks immediately for pollution. Brantley pointed out that groundwater data are difficult to obtain. She praised Pennsylvania, Colorado, and Texas for widely distributing their groundwater data and encouraged more states to put their data online so that the public can be better protected. Eric van Oort, University of Texas at Austin, remarked on the work on large datasets that Lackey presented, which could be useful both to prioritize wells for permanent abandonment and to locate wells with good integrity that could be reused and reprioritized. He proposed that academia and governmental organizations collaborate to entice operators to share more of their data to further this work. He also mentioned work at Heriot-Watt University on probabilistic modeling, which could allow projections about which wells are likely to leak and at what severity. He said that adding this approach could help develop a tool to assist with future forecasting.

6

Remediation, Reclamation, and Restoration

INTRODUCTION

The fifth session of the workshop, moderated by workshop planning committee member Mary Kang, McGill University, highlighted approaches to remediation, reclamation, and restoration as well as associated challenges. She described "remediation," which is a key step before reclamation that either follows or is part of the plugging process, as the clean-up of hazardous substances related to the removal, treatment, and containment of pollution or contaminants from environmental media such as soil, groundwater, or sediment. She pointed out that the terms "reclamation" and "restoration" are sometimes used interchangeably to represent the step that follows plugging. The Bureau of Land Management (BLM) defines reclamation as follows:

> Reclamation helps to ensure that any effects of oil and gas development on the land and on other resources and uses are not permanent. The ultimate objective of reclamation is ecosystem restoration, including restoration of any natural vegetation, hydrology, and wildlife habitats affected by surface disturbances from construction and operating activities at an oil and gas site. In most cases, this means a condition equal to or closely approximating that which existed before the land was disturbed. (BLM, n.d.)

Kang explained that because "primary succession"—that is, what nature will do on its own without human assistance—can take 1–10,000 years, "assisting" that process can be valuable. She suggested that the particular type of land cover at well sites influences the approach to the reclamation process: forest and agriculture are the top two types of land cover surrounding documented U.S. orphaned wells (Kang et al., 2023). She added that reclamation is expensive; however, the benefits might outweigh the costs—for example, the immediate cost might be ~$6.9 billion, while the value of agricultural benefits and carbon sequestration realized over 50 years from restoring oil and gas well infrastructure might be ~$21.3 billion (in 2018 dollars; see Haden Chomphosy et al., 2021).

Kang invited the session's speakers to discuss the following questions: How do remediation and restoration options vary among sites? What are the key metrics (e.g., vegetation, wildlife, ecosystem services, fragmentation)? What long-term maintenance and monitoring efforts are needed? What data gaps exist, and how can new technologies address them?

NATIONAL PARK SERVICE ORPHANED WELLS PROJECTS

Forrest Smith, National Park Service (NPS), highlighted the challenges that the national parks confront, with ~2,600 oil and gas wells located across ~55 park units. He indicated that 1,200 of these wells have been inspected and inventoried; at least 45 orphaned wells have been documented, and more are expected to be found as inspections continue. More than 100 of the wells that have been inspected—some of which had been plugged previously—are leaking methane.

Smith provided three examples of completed NPS reclamation efforts. First, Ohio's Cuyahoga Valley National Park, an urban park located between Cleveland and Akron, has 150–160 oil wells. The FF Hunt oil well was located near the base of a ski hill and had been orphaned for ~30 years; the topsoil was fouled from hydrocarbon and produced water spills and had to be removed and replaced. To reclaim this site, he explained that trees were cut down, the well was plugged, contaminated soil was scraped off, and a mix and till refreshed the soil. The site was then recontoured, the trees were brought back, and the site was reseeded and re-mulched. Another well in Cuyahoga Valley National Park, the EC Bender oil well, was partially plugged and orphaned ~80 years ago. Although NPS did not find soil or water contamination, the oilfield debris was significant. After the well was plugged, the site reclamation plan was similar to that for FF Hunt. He noted that each national park in the United States has unique documentation and goals for restoration, which is challenging; for example, Cuyahoga Valley National Park chooses to restore in a way that represents the park circa 1850.

Second, Smith said that in West Virginia's Gauley River National Recreation Area, the Mower Lumber well was orphaned for more than 30 years and was leaking methane before being plugged. During plugging, NPS found that the flow line was connected to other active wells farther down the mountain, and the operators of those wells were notified after the flow line was plugged. Old mountain roads to the site were then recontoured, trees were cleared, and vegetation mulch was placed.

Third, Smith noted that in the Big Thicket National Preserve in Texas, the Arco Rafferty well site and the Arco Rafferty common tank battery both had soils contaminated significantly with heavy metals (e.g., chromium, barium, and lead). The remediation plan included removing 6 inches of contaminated soil, bringing in native soil to replace the contaminated soil, doing a mix and till, and collecting more soil samples, and the sites are now below the allowable levels for heavy metals.

Smith then discussed four NPS sites that likely will be reclaimed next. First, in the Guadalupe Mountains National Park in Texas, the Pure Oil well was drilled in the 1920s and deepened in the 1940s; the oil company left the well since it had little economic value, and local farmers filled it with sediment and mud and produced water for their cattle. In the 1970s, the farmers then left the well after electrical lines to the site were destroyed in a storm. The site thus includes the original pump jack, a derrick frame that has fallen into the well site, an old tractor, tanks, and ample debris—all of which are hazardous to visitors and wildlife. Furthermore, the site is very difficult to access, with the need for mule trains, helicopters, and a 10-hour hike. He pointed out that after 50 years, items in national parks become antiquities that have to be preserved; therefore, this reclamation plan includes only removing the dangerous piles of debris and stabilizing the equipment.

Second, Smith described the Gherini well located in California's Channel Islands National Park that was drilled in 1965. It was partially plugged but completely abandoned after a workover failure in the 1970s. However, the site is now an antiquity and archaeologically sensitive owing to native burial sites, so NPS is working with local tribes to avoid disturbing these sites during future reclamation efforts. Third, he stated that Jean Lafitte National Historic Park in Louisiana has many wells suspected to be unplugged. Because boat traffic is being impeded and contact could cause spills, wellheads, pier structures, and flowlines likely will be removed.

Finally, Smith mentioned that NPS also has funding to plug six orphaned wells that are actively leaking methane in the Big South Fork National Recreation Area in Tennessee and Kentucky. He pointed out that all of these wells were previously plugged and abandoned in the past 20–50 years. If funding is secured, an additional six wells could be plugged in 2025.

SURFACE RECLAMATION AND RESTORATION

Ron Krawczyk, Parsons Corporation, described his work in Michigan to clean up more than 1,000 legacy oilfield sites across four counties since 2006; in Canada to conduct more than 800 methane studies on well, subsurface, and surface casing vent flow; and in New York to manage pre–Infrastructure Investment and Jobs Act (IIJA) funds and to plug more than 80 high-risk orphaned wells since 2018.

Krawczyk elaborated on the Parsons Corporation's Orphaned & Legacy Well Programs, which locates and researches wells; evaluates risk factors; identifies the highest-priority wells to address; and identifies low-risk wells for lower-cost, more passive approaches. Parsons also conducts environmental investigations, re-engineering for access, and plug and abandonment work. To locate wells, he explained that Parsons leverages ground- and aerial-based magnetometry; the latter has been particularly helpful for cases in which the historical record locations are far from the actual site. In many cases, once the well is found, it has to be rebuilt before it can be plugged. He noted that Parsons also does turn-key orphaned well-plugging projects with private corporations and state agencies in Arizona, Michigan, New York, Texas, and Canada.

In addition to plugging and abandonment projects, Krawczyk said that Parsons conducts monitoring for the Michigan Department of the Environment, Great Lakes, and Energy's Orphaned Well Methane Monitoring program with remote methane leak detectors and high-flow instruments, which are all tunable diode laser absorption spectroscopy–based. Most leaks that Parsons confronts are small-scale (less than 20 g/hour), and he noted that the company has achieved field results of leak quantification down to 0.3 g/hour (nondetectable).

Krawczyk indicated that Parsons also has engaged in research and development on risk-based alternatives to plugging wells with complicated re-entry or those that are only leaking methane to the surface without other fluid impacts. For example, Parsons has patented a methane biofilter that can be placed on top of a well to enhance the biodegradation of methane (e.g., converting methane to carbon dioxide) and to prolong the time until plugging can be completed.

Furthermore, Krawczyk continued, Parsons participates in many remediation and decommissioning activities for oil, gas, and other industrial contaminants. For example, in Michigan, Parsons operates a landfarm facility where it treats 100,000–120,000 cubic yards per year of petroleum and salt-impacted soil for reuse as backfill on well sites; since the facility was built in 2006, more than 1.5 million cubic yards have been remediated. Parsons also removes World War II–era asbestos flowlines and operates a Class II underground injection control well for leachate disposal.

Krawczyk emphasized that many of the challenges associated with remediation and reclamation of legacy orphaned well sites relate to access issues. For example, early wells might be located next to rivers and streams or were drilled in uniform spacing patterns where the nearest point of solid ground was selected for the well. Krawczyk also noted that some wells, particularly those located in swampy areas, have small well pads that require additional engineering and permitting for a service rig and have impassable entry roads. Furthermore, he commented that access might be limited owing to hunting seasons, agricultural planning seasons, wet seasons, nesting and migration, fish spawning, endangered species habitats, forestry diseases, cultural areas, and seasonal recreation. Another challenge relates to the number of entities from both private and public lands that would ideally be engaged. Special considerations to affected parties include utility right-of-way, corporation-owned property, municipality-owned properties and roadways, and disinterested or adverse property owners. Yet another challenge relates to land use changes between the time the well was drilled and present-day—rural land might now be developed heavily, dry land might now be a swamp or a lake, and a clear-cut area might now be densely forested (see Figure 6-1).

He observed that modern infrastructure (e.g., electric, water, gas, sewers, and buildings) as well as oilfield redevelopment might create additional land use challenges that involve increased coordination with other parties. Many levels of permitting might be required, he continued, such as with the U.S. Army Corps of Engineers, the federal government, state agencies, county agencies, and the Fish and Wildlife Service. These permits often drive restoration requirements, such as for certain seed mixes, soil erosion controls, finish grade evaluations, access roads, and other infrastructure. To minimize disturbance to communities and increase efficiency while working on orphaned well projects, he highlighted the use of early desktop reviews to categorize sites by complexity and estimate lead time for surveying, early engagement with affected parties, up-front drone surveys to pinpoint wellbore location and former access roads, and phased work queues.

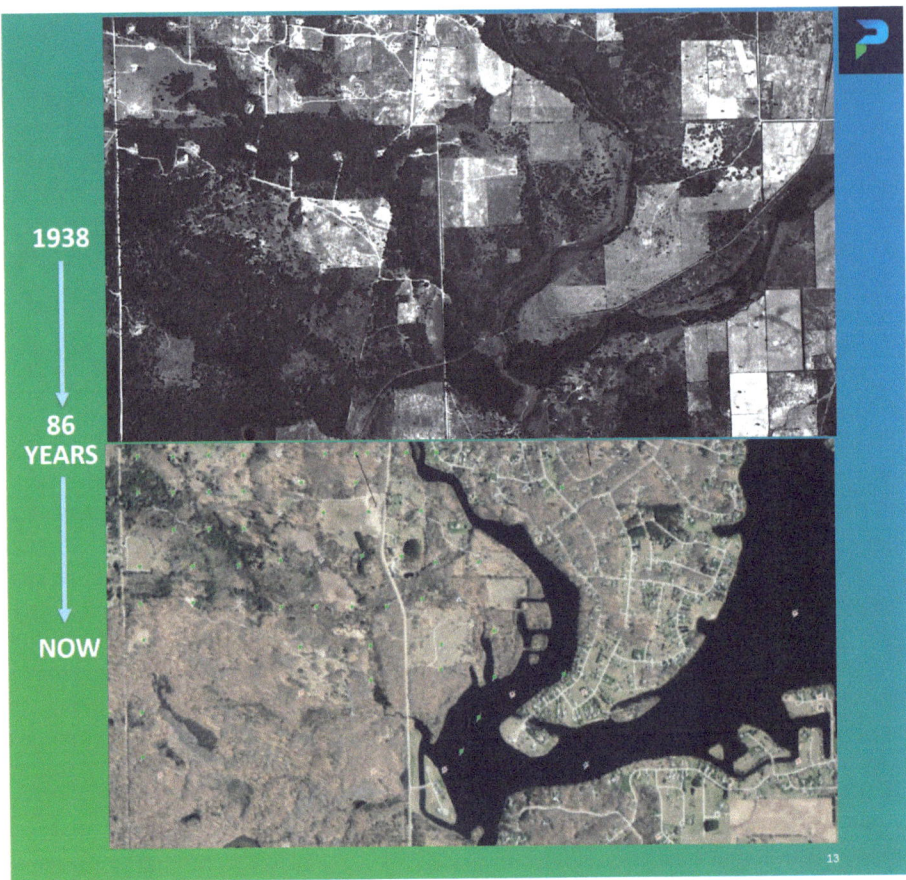

FIGURE 6-1 Drastic land use changes over a period of 86 years.
SOURCE: Krawczyk, 2024; Michigan State University RS&GIS Aerial Imagery Archive (upper photo); State of Michigan Department of Environment, Great Lakes, and Energy GeoWebFace Data Viewer (lower photo).

COLORADO'S ORPHANED WELL PROGRAM

Michael Hickey, Colorado Energy and Carbon Management Commission (ECMC),[1] emphasized that the ECMC has embraced the IIJA's suggestion to engage family-owned businesses. The businesses with which the ECMC partners have made significant adjustments, meet the new federal requirements for monitoring, data collection, and reporting.

Hickey explained that before accessing a site for plugging and abandonment and digging up well casing in Colorado, a remediation plan (Form 27) must be reviewed and approved. Thus, the planning process begins weeks to months ahead of the work. The

---

[1] For information about the ECMC's Orphaned Well Program and its annual reports, see https://sites.google.com/state.co.us/cogcc-owp.

remediation plan requires that once the wellhead is removed, confirmation samples will be submitted to verify that the dirt being left behind is clean. Additionally, a Notice of Intent to Abandon (Form 6), which describes plans for how the well will be plugged and includes a wellbore diagram, has to be reviewed and approved by the ECMC engineering group. He indicated that collaboration across the ECMC helps facilitate a smooth approval process. Form 42 is then distributed 48 hours before the rig is placed at the site, a second version of which notifies the local government designee that a well will be plugged in that person's jurisdiction. After work is finished, a Form 6 Subsequent Report needs to be submitted to detail the actual plugging work that was completed.

Hickey conveyed that the ECMC's pre-plugging methane monitoring is conducted using a variation of the chamber measurement approach that was described in the workshop's previous session. He said that combining methods by using an opened-up, static-free bag to form a background on the upwind side of a well to minimize wind interference gives more accurate high-flow measurements. When optical gas imaging is used, that same bag provides a clean background on the infrared images, which also helps to improve the accuracy of the measurements.

As a result of Colorado's new regulations, Hickey noted that the ECMC also is actively removing miles of flow lines. Several issues arise during this process, including that the lines cross several types of public and private properties and surfaces, are corroded, are more numerous than expected, and/or are filled with fluid and result in large clean-ups. When removing these flow lines, if an incidental spill is found, he stated that contractors are encouraged to clean up as much as possible. Furthermore, third-party professionals conduct ongoing monitoring and sampling throughout this work; photo ion detectors uncover hydrocarbons effectively, but inorganics are more difficult to detect and control.

When confirmation samples are positive and/or when messes are significant, Hickey underscored that remediations are difficult to schedule and to finish because repeat visits are often needed. He provided an example of the challenges of remediating a well site in which gas condensate had been leaking into the ground for 30 years, extending beyond ECMC's initial perimeter and threatening domestic wells and adjacent homes. Sharing some of the maps and volumetrics of this site that were developed by a drone operator, he noted that while some of the soil that was removed was determined to be clean enough to be reused as backfill, 3,555 cubic yards of impacted soil were transported for disposal. He estimated that excavating and stockpiling the overburden; excavating, loading, transporting, and disposing of impacted soils; purchasing, loading, and transporting fill; and replacing the overburden and reclaiming the site cost $100 per cubic yard of impacted soil—a very expensive undertaking. He explained that after the remediation of this site was compete, reclamation experts performed soil treatments on the surface so that homes could be built on the site.

A MODERN APPROACH TO RECLAMATION

Brent Wilson, Red Willow Production Company, explained that the Southern Ute Indian Reservation comprises an area of 1,058 square miles in southwestern Colorado.

Red Willow Production Company was formed in 1992 to take ownership of the tribe's energy resources within the San Juan Basin. Since then, it has expanded to include reserves and production in the Deepwater Gulf of Mexico and in the Green River and Delaware Basins.

Wilson indicated that state and local laws do not apply to the Tribe's operations occurring within the Southern Ute Indian Reservation, and Red Willow's operations within the exterior boundaries of the reservation are conducted under tribal and federal regulatory oversight. Wells on tribal leases are overseen by the Southern Ute Department of Energy, the Southern Ute Department of Natural Resources, BLM, and the Bureau of Indian Affairs (BIA). He noted that reclamation projects are considered complete when the wells are released through the approval of Final Abandonment Notifications (FANs), after they meet the criteria of and pass inspection by the Southern Ute Department of Energy and BLM. This situation is unique, he continued, in that Red Willow (which is owned by the Southern Ute Indian Tribe) is the oil and gas operator and the regulatory body that works with BLM and BIA, and the tribe is also the landowner; this relationship has allowed Red Willow to prioritize reclamation practices and give the land back to tribe members to use as they wish.

Wilson stated that the reservation is located in a high-desert climate. Severe drought conditions have persisted for at least the past 10 years, creating challenges for reclamation with less spring moisture available to support germination. Red Willow aims to build an idled well program that accounts for and mitigates these drought and climate issues, which make reclamation a priority. He remarked that succession is the main goal in reclamation. The process begins with bare ground and annual weeds, and then moves to establishing perennial grasses and shrubs over a period of 2 to 4 years; from that point, he said that Mother Nature can establish a forest over 5 to 20 years. He encouraged states to consider during their planning processes how much time and effort are required for reclamation.

In 2019, Red Willow created an internal team to focus on the increasing number of idled wells and to develop new processes and techniques to reclaim locations efficiently and reduce plugging and abandonment liability, Wilson explained. This team leverages data to develop reclamation plans that include contour maps, seeding and fencing details, stormwater management, soil analytics, and prescribed application rate for soil amendments and mulch (see Figure 6-2). After generating these plans, the team communicates with the Southern Ute Department of Energy and tries to accomplish reclamation projects in a single attempt instead of having to return over the years to establish growth.

To capture some of the detailed topographic data included in these reclamation plans, Wilson indicated that Red Willow leverages consumer-level drone-based LiDAR technology. Drone photos are used to generate a three-dimensional model in AutoCAD, which is then manipulated to determine a final grading plan. The goal is to select contour lines and find their counterparts on the other side of the well pad, and then draw them to create a consistent flow of surface through the well pad. This maximizes the amount of time that the water can stay on the well pad without turning into a pond, encouraging a gentle drainage of the location. He emphasized that this preliminary work ensures

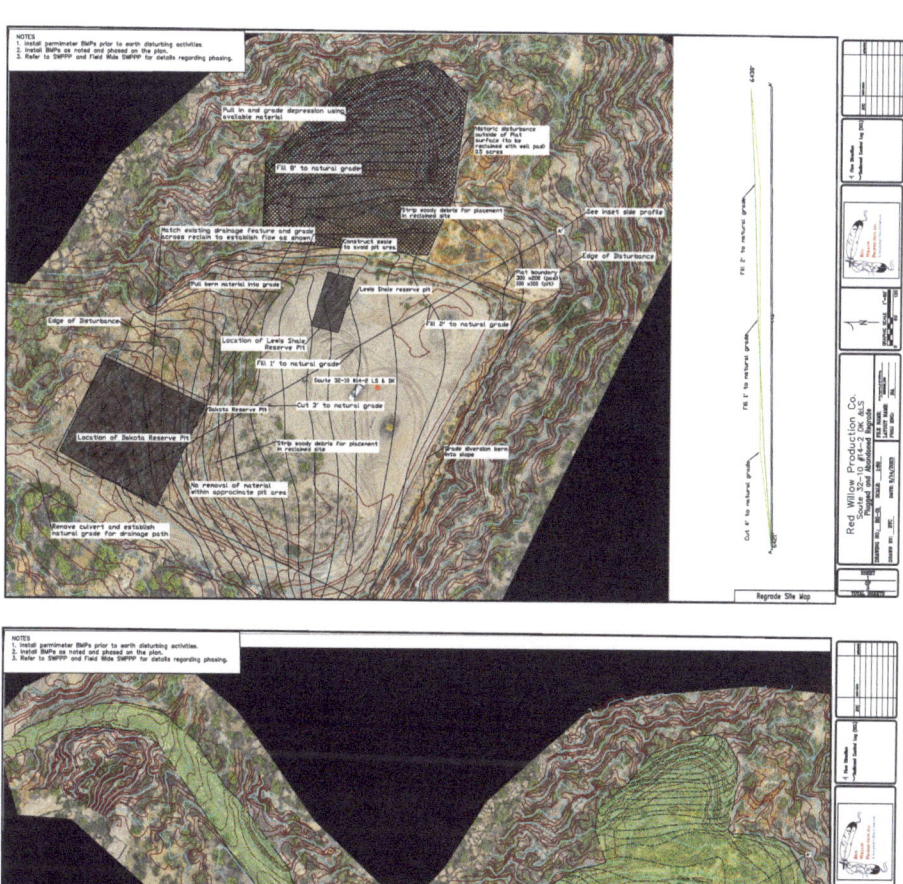

FIGURE 6-2 An engineered approach to reclamation.
SOURCE: Wilson, 2024.

that something stable promotes succession, grows the species, and helps reclaim the area. Drone cameras are also used to monitor vegetation and track the success of the reclamation over time.

Wilson added that Red Willow has been working with several soil additives and new preparation techniques to help establish initial vegetation. For example, biosolids, which are by-products of the tribe's wastewater treatment plant, are used to introduce nutrients back into the sterile soil. Biochar, produced from beetle kill trees in Colorado, absorbs contaminants, helps neutralize soil chemistry, and increases soil water retention. Surface soil flipping, a technique in which the top foot of soil is flipped, prevents ground sterilant remnants from weed spraying from directly contacting the seed bed. Furthermore, after realizing that straw mulch is ineffective in a windy climate with little rain, he said that Red Willow adopted the use of excelsior mulch for shallow slopes—which is difficult to apply but traps moisture in soil—and hydromulch for steep slopes—which is expensive but easy to apply.

In closing, Wilson noted that Red Willow and the Southern Ute Indian Tribe have experienced marked improvement in reclamation success as well as a reduction in time to receive FAN approval. The team strives to continue to identify and evaluate new technologies and techniques to drive progress on site reclamation.

OPEN DISCUSSION

Kang moderated a discussion among the workshop speakers and participants. She inquired as to how costs per well change when restoration is a high priority. Wilson reiterated that Red Willow prioritizes returning surfaces to the tribe so that they are able to hunt. He said that Red Willow budgets twice the cost of plugging for a reclamation project, but that amount does not include soil remediation. Wilson noted that about 1 in 30 remediations are a major effort. Smith noted that NPS also budgets heavily for reclamation, with the goal of achieving a nearly "pre-operation state" of the area. He added that this budget often depends on the difficulty of gaining access to sites. In some cases, surface cost can be double the downhole costs.

Greg Lackey, National Energy Technology Laboratory, asked about the proportion of well pads from which significant amounts of soil are being removed for remediation. Hickey replied that in Colorado, large sites like the one he described during his presentation have only been found a few times over the past couple of years; however, predicting where they will occur is difficult. Krawczyk said that one-half to three-quarters of the wells that his team has remediated have had significant amounts of soil removed, in part owing to the fact that many of the wells were drilled in the 1920s. He observed that having good site investigation, analytical samples, and environmental consultants is key to mitigating risks.

Adam Peltz, Environmental Defense Fund, observed that when salt is found in soil, it must be collected and discarded. He asked if any strategies exist to address groundwater contamination, in particular saltwater intrusion into a freshwater aquifer. Krawczyk agreed that if brine is left in the soil, reclamation will fail. In terms of the saltwater intrusion issue, he said that three-dimensional modeling techniques could help deepen understanding of the problem. Hickey remarked that for an area of 38 acres where salt intrusion has occurred over the course of 50 years, his team is investigating the possibility of reusing an orphaned well that was designated as an injector: fresh water would be used to flush the salt kill, and the water would be collected and then reinjected. Krawczyk added that for those who have disposal wells, a pump and treat system could be used and old oilfield infrastructure could help remediate.

Dan Arthur, ALL Consulting, referenced the Department of Energy/National Energy Technology Laboratory's research on brine-impacted soil clean-up using phytoremediation, soil amendments, and managed irrigation with produced water. He explained that the choice of solution depends on how quickly clean-up has to be completed; if the clean-up process could span a few years, more options are available. Krawczyk mentioned that phytoremediation has been effective for crude oil compounds. He added that in situ remediations often prompt more stringent regulations, owing to a concern about heavy metals and daughter compounds.

7

Advances in Plugging and Abandonment for Idled Wells

INTRODUCTION

In the final session of the workshop, planning committee member and moderator Mileva Radonjic, Oklahoma State University, invited speakers to discuss advanced technologies for orphaned and idled wells. The goal of the session was to provide an overview of the industry's highest standards for plugging and abandonment and to begin to break existing silos across disciplines and communities.

LESSONS LEARNED FROM THE AMERICAN ASSOCIATION OF PETROLEUM GEOLOGISTS: 2020–PRESENT

Susan Nash, American Association of Petroleum Geologists (AAPG), said that AAPG's Division of Environmental Geology aims to preserve the integrity of the environment, especially with consideration for oil and gas operations and issues related to groundwater, surface water, and induced seismology. She stressed that this mission can be achieved by communicating, sharing information, and collaborating across disciplines as well as by leveraging new technologies such as artificial intelligence (AI), large language models, and analytics.

Nash explained that AAPG shares information via workshops, webinars, research conferences, an annual convention, and publications such as *AAPG Bulletin* and *AAPG Explorer*. It also hosts platforms, for which AAPG membership is not required, that encourage discussion and collaboration. On the topic of orphaned and abandoned wells, AAPG has organized 10 webinars (e.g., on methane detection, modeling the subsurface, modeling cause and effect) since 2020 and several workshops in cities across the United States since 2022. The next workshop is scheduled for February/March 2025. These events extend beyond discussions of current regulations to include conversations about strategies to tackle the problem; she reiterated that knowledge sharing across the states is key. New insights that states gather during these events relate to transparent data sharing, best practices, consistency, proactive and pre-emptive approaches for marginal wells (i.e., future orphaned wells), opportunities for both traditional (e.g., bonds) and innovative (e.g., carbon credits) funding sources, data integration and analytics (e.g., AI and machine learning), and repurposing.

Nash noted that platforms for knowledge sharing and support continue to expand across the field. For example, the U.S. Geological Survey (USGS) is developing databases to inform work with orphaned wells, abandoned wells, and remediation, and the Department of Energy is developing an initiative to help locate undocumented orphaned wells. Furthermore, AAPG's Technology Showcases highlight new technologies (e.g., sensors for methane detection, monitoring, and measurement; materials for plugging; new analytics) and help people to connect with those working on the ground, to consider how to fund new technologies, and to develop new partnerships with technology companies to plug wells more efficiently and effectively.

Nash encouraged "reconfiguring our imagination" to think about data and problem-solving in new ways (e.g., how knowledge about modeling fluid flow at the subsurface can be used to understand how other wells could be affected and how new orphaned wells could be identified). She underscored that orphaned wells present far more than just a technology problem. Orphaned wells represent an opportunity for repurposing to augment the nation's water and energy supply, as well as serving as potentially novel ways to expand the nation's infrastructure.

NEW AND UPCOMING PLUGGING AND ABANDONMENT TECHNOLOGY

Thomas Lopez, ExxonMobil, noted that in 2024 alone, ExxonMobil will complete 2,000 plugging and abandonment projects globally. Before providing a general overview of emerging technologies, he underscored that rock-to-rock barriers are required to contain sources of inflow, and ExxonMobil's principal objective is restoration of the caprock. He explained that barriers are meant to be permanent; thus, it is important to use proper materials that will not degrade over time, incorporate a long enough plug, be placed properly in an impermeable caprock with sufficient strength to withstand future recharge pressures, and meet regulatory and technical requirements.

Lopez mentioned that although ExxonMobil primarily focuses on offshore operations, many of its philosophies still apply to onshore operations. He encouraged exploring novel execution techniques (e.g., "through-tubing" cement plug placement) and avoiding activities that are too costly, complex, or time-consuming and do not improve long-term well integrity. He proposed that equipment always be "right-sized" (i.e., smallest-footprint equipment spread) and cost-effective for a job, and he suggested using a phased plugging and abandonment approach by grouping similar-phase tasks on different wells with the same equipment.

Lopez indicated that ExxonMobil prioritizes rigless plugging and abandonment in its offshore work to reduce equipment spread. Rather than pulling tubing and performing operations on the drill pipe, they leave the equipment in place, set mechanical plugs and punch tubing, and place cement plugs from the surface. If contamination is a concern, longer plugs can be used.

Another concern in offshore operations relates to poor annular cement, Lopez explained. Using a contemporary approach, one might perforate and circulate cement around the backside or use section milling. Modern alternatives to these approaches include the use of (1) "perforate, wash, and cement," in which holes are punched and cement is pumped across the interval; and (2) crept shale, in which one identifies, modifies, and hydraulically tests existing ductile shales that have formed a natural barrier around an exposed casing. New and developing technologies include casing ablation techniques (essentially vaporizing the casing using propellant) and a coil tubing erosional technique.

Cement bond logging tools, Lopez continued, are also in the process of being improved. Traditional versions of these tools require the removal of tubing to evaluate the cement behind the casing. However, newer versions operate "through-tubing" to discover what is behind the casing without removing the tubing. He pointed out that this technology is not yet mature; although it uses some machine learning, one would need to account for the exact combination of casing sizes, tubing sizes, weights, cement grades, and formation slowness. He also mentioned that this technology could enable work with less equipment on site in both the onshore and offshore environments.

Lopez next briefly highlighted a few alternate abandonment materials—for example, bismuth alloy plugs and thermite-based plugs. A thermite tool, which reaches 2,500 °C in a wellbore, can be used to melt existing tubing, residual cement, casing, and formation. However, although it forms a type of plug, he said that this plug is not gas-tight and thus is not providing a seal. As a result, some companies are using the thermite plug first and then using a corrosion-resistant bismuth plug on top of it to provide a gas-tight seal. He noted that the latter option creates a very small plug with only a few feet of cement, which could be problematic if issues exist with the caprock.

Lopez explained that other companies are focused on addressing issues with leaking channels. Approaches include biomineralization (i.e., pumping in bacteria to convert salts in the fluid to a solid plug), mineral scale (i.e., pumping in incompatible waters to produce mineral scale in small spaces), dissolvable glass (i.e., dissolving glass into a stable solution, which precipitates on contact with cement, rock, or steel), and mechanical cement densification (i.e., creating dents in the casing that expand to compact the cement).

In closing, Lopez emphasized that many new approaches to plugging and abandoning wells are emerging, but barrier quality remains a top concern; new techniques have to be as good as or better than traditional techniques if they are to be deployed. Subsequent goals relate to continual gains in efficiency and safety; if efficiency improves by finding the right equipment and the right solutions, more wells can be plugged. Instead of thinking about others in the abandonment space as "competitors," he encouraged cooperation and learning from one another's experiences. He urged companies, universities, and governments to work together to support, promote, and test emerging technologies.

URBAN ABANDONMENT: CHALLENGES AND SOLUTIONS

Jesse Frederick, WZI, emphasized that location is key when selecting approaches for abandoning wells. Challenging logistical issues arise in urban areas in particular; for example, in Los Angeles, orphaned wells have been discovered on the properties of urban high schools and churches as well as when residents prepare to install pools or add basements to their homes. In addition to residential areas, he said that urban orphaned wells might be found in commercial areas, open areas, existing oilfields, and abandoned oilfields.

Frederick indicated that when these wells are discovered, several issues could arise relating to pressure recovery, bad or missing casing, missing well history, cement, equipment, project control, or contracts. He noted that a well in a recovering field is particularly dangerous owing to pressure recovery issues; if the well has a wellhead but no valves or gauges, one could penetrate a pressure barrier (i.e., encounter unexpectedly high pressures in the wellhead). Over the past two decades, industry professionals have learned that just because one has finished pumping a well does not mean that it has no pressure or will stay depressurized; as long as fluids are left in the reservoir that reflect the original composition, a new pressure will emerge and rebalance. Thus, he stressed that knowing the conditions of wells before entering them is critical. An important first step before contracting for work on an orphaned well is likely conducting "orphaned well triage," which helps to highlight potential issues and determine priority based on achieving key objectives (e.g., whether the goal of a project is to reduce methane or to address an acute hazardous issue).

Elaborating on the potential issues and risks that contractors who plug and abandon wells face, Frederick explained that on the way into a well, problems could arise while removing the top seal, drilling out the first plug, finding fish, encountering loss of circulation, or encountering casing failures (e.g., casing corrosion, casing that was shot off and pulled, damaged casing, casing filled with debris). Tools to address these casing issues include active and passive ranging, drill collar severance, and concrete cure/temperature logging. On the way out of the well, issues could arise from a poor cement job or poor seals in the upper intervals. Well-to-well communication is a particularly challenging problem to address, he continued. When Los Angeles first became more urban, people were encouraged to put wells into a common well cellar; however, when wells are clustered, they start to communicate with each other as they age (i.e., near the surface where they converge into the common cellar–not in the reservoir). Frederick and his team developed a way to identify which wells are communicating, which is key for accurate pressure monitoring during well abandonment.

Turning to a discussion of various plugging materials, Frederick distributed several samples to workshop participants, including compressed bentonite (which expands to 250% of its volume and has low permeability), Class G cement, and novolac (a thermosetting bisphenol plastic). He noted that novolac can be used to seal micro-annuli, although it is difficult to manage and pressure control is essential. He encouraged the use of ample sand to act as heatsink when using novolac, given its high temperature; then, in combination with nanoparticles, it will form a firm bond to steel and provide a "superior" seal (see Figure 7-1). He added that sodium silicate does not achieve success-

Figure 2.5-A
22 Jun 8:05 am

Figure 2.5-C
25 Jun 9:30am

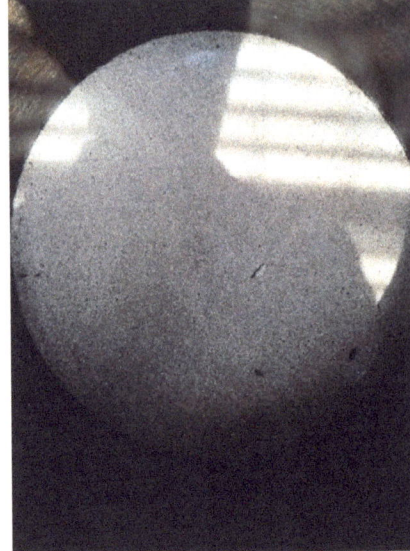

Figure 2.5-B
22 Jun 8:05 am

FIGURE 7-1 Novolac cures quickly and serves as a superior sealing material.
SOURCE: Frederick, 2024.

ful seals of micro-annuli. Furthermore, he agreed with the concerns expressed earlier in the workshop about inferior cement quality. For example, issues with cement could include a poor cure, channeling and micro-annuli, and leakage. Additives can lead to a better outcome, however, and he suggested latex as the best additive for cement to preclude gas.

Frederick next offered several suggestions to enhance project control for plugging and abandonment jobs. He suggested that dedicating funding and effort to sealing the top of the productive hydrocarbon layer could be beneficial. He also suggested the American Petroleum Institute (API) could release a series of recommended practices for wells of various types, which could be used by the private and regulatory sectors. This type of support could help to ensure that people have met regulatory and technical requirements and could be used as an instrument in the future to assign emissions reduction credits. He underscored the value of preparing back up plans for all plugging and abandonment jobs. Looking forward, he mentioned five additional suggestions for the industry: (1) create a "greenhouse gas bounty" on orphaned wells, (2) establish a rational requirement for completion of plugging and abandonment, (3) do not trust casing because it corrodes, (4) establish rational criteria for liability protections to

U.S. GEOLOGICAL SURVEY SCIENCE TO SUPPORT ORPHANED WELL-PLUGGING: HISTORICAL DRILLING, PRODUCED WATERS GEOCHEMISTRY, AND GROUNDWATER QUALITY

attract experienced teams (e.g., a temporary federal agency similar to the Resolution Trust Corporation to absorb liabilities of wells, if all criteria are met), and (5) establish equipment and qualification criteria for a site early in the process.

Nick Gianoutsos, USGS, provided an overview of oil and gas drilling in the United States, noting that the first commercial oil well was drilled in Pennsylvania in 1859, leaving a legacy of more than 160 years of drilling and abandonment. He indicated that plugging with cement became standard in the 1950s and wells abandoned before then likely were plugged insufficiently and are now "orphaned." He defined orphaned wells as "unplugged, nonproducing oil and gas wells that have no responsible owner or party to remediate the well site, leaving the burden of plugging and reclamation to the local, state, or federal governments, and occasionally to landowners and nongovernmental organizations." Recalling discussions from the first day of the workshop, he added that the United States now has nearly 150,000 documented orphaned wells and 250,000–750,000 undocumented wells, and according to one study by Raimi and colleagues (2021), the median cost of plugging each well is $76,000—an amount that varies by well age, depth, condition, and cement type.

Gianoutsos explained that USGS has created a series of products that can be used to enhance decision-making for plugging orphaned wells. In 2022, USGS released the first publicly available national-scale dataset (Grove and Merrill, 2022), including more than 117,000 orphaned wells across 27 states. The orphaned oil and gas well dataset includes API numbers and latitude and longitude coordinates. This dataset represents a snapshot in time as the population of orphaned wells continues to evolve as more wells are documented and other wells are plugged, but it remains a useful analytical tool, he said. For example, analysis reveals that when energy was in high demand and oil prices were high (e.g., in the 1940s and 1980s), more wells were drilled and a higher percentage of those wells became orphaned. Also, as technology has improved over time, well depth has increased. Currently, the deepest well in the dataset is 23,431 feet deep, drilled in the 1980s in Oklahoma. He added that vertical wells comprise 98% of the dataset (Merrill et al., 2023).

Gianoutsos noted that if a well plug is expected to last thousands of years, understanding the environment in which the plug will be placed is critical. He suggested use of the USGS Produced Waters Geochemical Database, which is available online and contains more than 100,000 samples of produced waters across the United States compiled over decades from published sources and from USGS sampling and analysis (Blondes et al., 2023). Samples include constituents and total dissolved solids as well as other information. He stated that the database's user interface allows users to search for the area in which they will be working or for a specific formation. The database also contains critical information about other minerals, which could inform the repurposing process for wells.

Another useful resource, Gianoutsos continued, is the USGS dataset on groundwater quality near orphaned wells, which was created in 2024 (Haase et al., 2024). Groundwater quality measurements within 1 mile or less of an orphaned well were collected from the USGS dataset of 117,000 orphaned wells across the United States and from the USGS National Water Information System. This USGS dataset on groundwater quality contains water quality parameters and indicators for corrosivity, for example. Initially, he said that USGS was most interested in the impacts that orphaned wells are having on groundwater quality, but now it is exploring the impact that groundwater quality is having on orphaned wells. For example, areas with high corrosivity may have wells that deteriorate faster and could pose a more immediate risk to groundwater.

In summary, Gianoutsos remarked that as drilling practices and techniques have improved, different sets of challenges have arisen for well-plugging based on the location, era, geology, and geochemistry of each orphaned well. Understanding these differences is key for decision-making, and he underscored that USGS's data products help enhance understanding of orphaned well emissions and inform efforts to plug orphaned wells.

ISOLATING ANNULI USING SHALE/SALT AS A BARRIER

Eric van Oort, University of Texas at Austin, presented an overview of RAPID (Rig Automation and Performance Improvement in Drilling), a consortium at the University of Texas at Austin that focuses on drilling and drilling automation, with a branch dedicated to well-plugging and abandonment and related well-integrity issues. He mentioned that RAPID is supported by industry; in particular, AkerBP, Equinor, and TotalEnergies have been strong proponents of the use of shale/salt as a barrier (SAAB).

van Oort pointed out that although cement is a good building material, it is not necessarily a good isolation material. In a comparison between ordinary Portland cement (OPC) and geopolymers (alkali-activated materials), he noted that OPC has higher compressive strength but lower tensile strength and significantly lower bond strength than geopolymers. When cement in used as a plugging material, he said that micro-annuli can form easily given this weak bond strength between the casing and the cement, causing a pathway for flow and cracks in the cement that will not heal. However, if geopolymers are damaged, they exhibit unique re-healing properties. As a result, some operators are replacing their OPC abandonment plugs with those made of geopolymers. However, he mentioned that it is unknown how stable and isolating either OPC or geopolymers will be over thousands of years, and accelerated aging tests cannot be conducted with cement.

Introducing a solution to this problem of unknown material durability, van Oort explained that 15 years ago, North Sea operators were reentering wells for abandonment, running cased-hole bond logs, and finding great bonding far above the top of cement. However, this phenomenon was only occurring in shale formations. The mechanism that enables this seal is creep—that is, a rock that behaves as a liquid and flows. Essentially, he indicated that the caprock is being replaced with actual caprock instead of an artificially introduced material with unknown behavior.

van Oort and his team began investigating this phenomenon for SAAB several years ago by building rock mechanics test setups (see Figure 7-2). They put cylindrical shale samples in a core holder, with a casing insert in the middle and an open annular space between them. The rock was confined, and downhole stresses and temperatures were applied. Perfect primary and secondary creep was observed filling in the annulus, forming a great barrier. When the barrier's permeability was tracked over time, he said that it eventually reached that of the native shale, which is 1,000 times less than that of cement, thus creating a barrier that is 1,000 times better than cement. A breakthrough test was performed next to verify the integrity of this barrier: the sample measured 3 inches and held 1,000 psi differential pressure. These barriers were found to be extremely strong: a shale section of only a few feet had the barrier power of a cement column of several hundred feet.

During continued laboratory observation of the barrier's behavior, van Oort indicated that the annulus closed over approximately 18 days. A finite element model was then constructed for simulation purposes, and parameters were extracted—in the field simulation, the annulus closed in approximately 90 days. However, he noted that barrier formation can be accelerated with thermal simulation—that is, placing a heater in the well to heat the casing and the formation behind it to accelerate the creep rate by orders of magnitude. He cautioned against heating to temperatures above 300°C, which would damage the shale.

van Oort explained that the types of North Sea shales that form these barriers naturally include geologically young shales with high clay content (with significant free and mixed-layered smectite), high porosity, high cation exchange capacity, low matrix cementation, low strength, low friction angle, and low compressional wave velocities. Whether North American shale/salt formations have these characteristics and display the same creep behavior remains unknown, and suitable candidates could be identified and tested in SAAB equipment. Furthermore, he and his team are preparing to run a new set of tests with larger samples for the second phase of SAAB experiments to determine if shale/salt creep can close micro-annuli and cracks in cement and re-establish annular pressure integrity. He shared that if an annulus is open and leaking methane, in the future one could stimulate the shale to form the annular barrier and set the abandonment plug (without the need for a rig) with geopolymers.

OPEN DISCUSSION

Radonjic moderated a discussion among the workshop speakers and participants. An online participant asked how USGS datasets are affected by the fact that spud dates for older wells were not always reported to or recorded by regulators. Gianoutsos responded that finding dates for the oldest wells in the United States is challenging, as drilling records started to become standard only in the 1950s. Thus, he said that the orphaned oil and gas well dataset was created by identifying wells with API numbers, gathering a spud or completion date from the S&P Global database, and mapping. Frederick suggested looking at archives of the *Oil and Gas Journal,* which dates back to 1902, for information on well activities categorized by location.

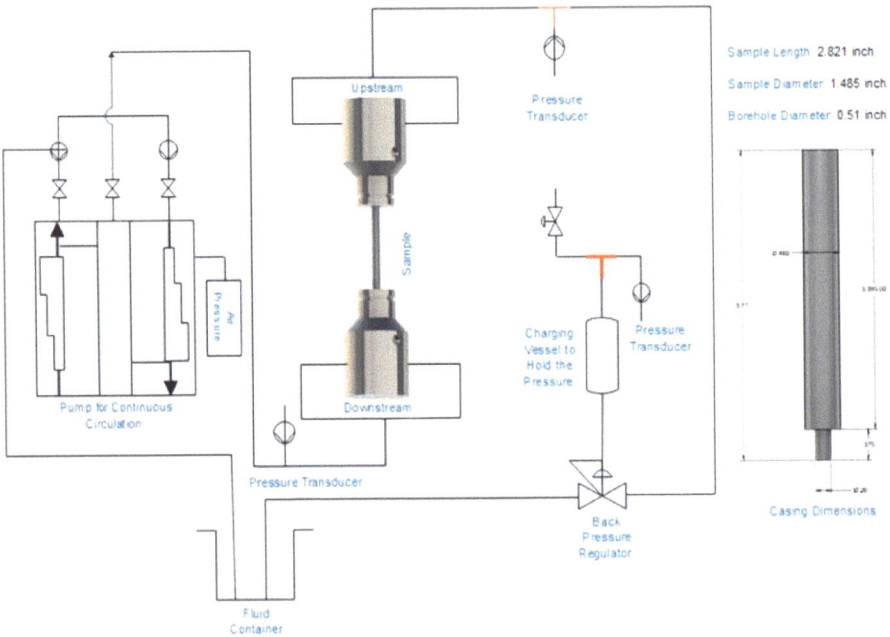

FIGURE 7-2 Schematic for the SAAB experimental setup.
SOURCE: van Oort, 2024.

Mohammad Khan, AHP Tech, inquired as to whether a correlation exists in how the wells included in the USGS dataset on groundwater quality lowered the alkalinity or increased the corrosivity of groundwater. Gianoutsos noted that making a direct correlation is difficult. His team is collecting groundwater quality measurements near the wells but cannot link the source of the constituents in the groundwater quality, especially when active wells surround orphaned wells. He emphasized that the primary goal of the dataset is for people to develop a basic understanding of the groundwater quality where they are working.

Khan also highlighted a disconnect between the plugging and abandonment industry and the cement and concrete industry, the latter of which has many technological advancements that could benefit the former. He encouraged increased collaboration between the plugging and abandonment industry and the American Concrete Institute. Nash mentioned that those conducting research on material strength have participated in AAPG conferences, and Radonjic expressed her support for an interdisciplinary approach to explore new materials and additives. van Oort reiterated that researchers are looking actively for alternatives to cement owing to its limitations; for example, an earthquake can shear a casing cement bond interface easily and create a leakage path. Frederick urged people to keep regulatory approvals in mind when considering the use of innovative materials; he encouraged more "dynamic" approaches in the regulatory sector as new technologies are introduced.

Greg Lackey, National Energy Technology Laboratory, asked what type of information would be useful for van Oort to develop confidence in North American shale barriers. van Oort replied that he would need to see the shale's properties and conduct SAAB tests, which are performed easily in the laboratory and use a standard protocol to determine whether the shale exhibits the appropriate behavior. Meg Coleman, Environmental Defense Fund, wondered if van Oort had considered the Bureau of Economic Geology as a source to test North American shales, but he responded that because some of their core samples are not well preserved and are dried out he has not pursued that path. In response to a request from Bryan McLellan, Alaska Oil and Gas Conservation Commission, van Oort encouraged operators to send core shale samples to his team for suitability analysis. Radonjic posed a question about the potential to use synthetic shale, but van Oort said that synthetic shale would not provide the degree of cementation seen in a real shale, and he would rather use field materials. He added that AkerBP currently is building shale barriers into their new well designs.

In response to a question from Nash, van Oort noted that distributed acoustic sensing networks have not been used to monitor for any micro-seismicity or other changes that could affect the integrity of shale barriers. Danny Sorrells, Railroad Commission of Texas, inquired about the temperature range applied for North Sea shales. van Oort explained that the optimum temperature to accelerate creep and form a better barrier is ~200°C, which can be achieved with a commercial heater. In theory, one could heat as high as 2,500°C with thermite, but that would be both unnecessary and detrimental to creep and re-healing properties.

An online participant posed a question about the role of petroleum landmen in finding, negotiating, and plugging orphaned and abandoned wells. Nash replied that even in cases in which an orphaned well is the responsibility of the state, the state might not own the minerals. This issue complicates the potential for carbon credits and who is entitled to them. A landman could help find both the title and the required surface owner permission as well as complete the mineral takeoffs and secure ownership. She highlighted the role of oil and gas attorneys in navigating these multi-ownership issues.

8

Examples of Key Workshop Themes

Workshop planning committee chair Mary Feeley, formerly of ExxonMobil, commended the workshop speakers and participants for interesting discussions that revealed the breadth and complexity of the issues associated with orphaned wells. She summarized examples of key themes from the workshop, including lessons learned, challenges, and opportunities mentioned by the speakers and participants over the course of the 2-day event. The highlighted themes signify discussions from the workshop, but are not statements of consensus.

Feeley described the impressive set of state experiences that was shared during the workshop. She reiterated that states have been doing this work for years, and understanding and sharing their plugging and abandonment best practices is critical to move the field forward—a possible next step is to consider how best to gather and disseminate this expertise.

Feeley highlighted similarities in guidance and practices across states as well as the ability of some states to adopt practices from other states. However, she said that states also have unique environments and conditions, and understanding complex geology (e.g., the subsurface, the movement of contamination away from the wellbore, and groundwater issues) is important for achieving successful plugging and remediation. Uncertainties in these environmental conditions are also complex; therefore, several workshop speakers observed that orphaned well programs benefit from the flexibility to plan and design for unique situations. They also benefit from having the ability and flexibility to address unforeseen issues that arise during the plugging and remediation processes.

Several speakers mentioned the importance of experienced and qualified professionals. Feeley emphasized speakers' assertion that the field could benefit from the inclusion and upskilling of more plugging professionals, regulators, inspectors, petroleum engineers, and hydrogeologists. Furthermore, it was mentioned in the workshop that well-trained operators are preferred for reentering wells, because re-pressurization, in particular, is challenging and risky. Many workshop participants noted that better assessing the qualifications of plugging and abandonment contractors could be beneficial for long-term plugging success, especially considering the number of wells that have had to be re-plugged.

Feeley next considered the workshop's focus on states' approaches to well-plugging prioritization, reflecting on their high-priority concerns about proximity to people, methane, surface water, and groundwater. In terms of the plugs themselves, several workshop speakers commented on the issues surrounding cement and noted that new and evolving technologies for barriers continue to emerge. As uncertainty about the lifetime of cement plugs remains, they encouraged continued research into alternate materials and approaches. Tagging of plugs was considered an important part of the plugging process by many speakers, and new guidance, regulations, and technologies (e.g., drones) for locating and monitoring wells continue to materialize.

Furthermore, reclamation, which is a top priority for tribal lands and parks, could be more successful if included in the plugging plan from the beginning, Feeley continued. Early planning for restoration of the soil and the surrounding environment to stop contamination issues from continuing after operations are complete is important. A few workshop speakers noted that reclamation might double plugging costs (i.e., the incremental costs of surface work could double the downhole cost, thus tripling the total); remediation is also costly but is an important step.

Feeley presented key uncertainties that remain, such as the best approach toward long-term monitoring to understand issues related to well integrity and groundwater contamination. Instead of monitoring every well, some participants suggested that targeted monitoring to better understand the issues of different aspects of plugging and abandonment could be valuable. Other uncertainties relate to the effects of microseismicity and considerations of regulations for well re-use and carbon credits. As the workshop drew to a close, she invited speakers and participants to continue to share their thoughts and provide feedback to move this work forward both in the upcoming National Academies of Sciences, Engineering, and Medicine consensus study on orphaned and abandoned wells and in the field more broadly.

References

Bachu, S. 2017. Analysis of gas leakage occurrence along wells in Alberta, Canada, from a GHG perspective – gas migration outside well casing. *International Journal of Greenhouse Gas Control* 61:146–154.

Barbot, E., N.S. Vidic, K.B. Gregory, and R.D. Vidic. 2013. Spatial and temporal correlation of water quality parameters of produced waters from Devonian-age shale following hydraulic fracturing. *Environmental Science & Technology* 47(6):2562–2569.

BLM (Bureau of Land Management). n.d. Oil and gas site reclamation. https://www.blm.gov/programs/energy-and-minerals/oil-and-gas/reclamation.

Blondes, M.S., K.J. Knierim, M.R. Croke, P.A. Freeman, C. Doolan, A.S. Herzberg, and J.L. Shelton. 2023. U.S. Geological Survey national produced waters geochemical database (ver. 3.0, December 2023). U.S. Geological Survey data release. https://doi.org/10.5066/P9DSRCZJ.

Boutot, J., A.S. Peltz, R. McVay, and M. Kang. 2022. Documented orphaned oil and gas wells across the United States. *Environmental Science & Technology* 56(20):14228–14236.

Bowman, L.V., K. El Hachem, and M. Kang. 2023. Methane emissions from abandoned oil and gas wells in Alberta and Saskatchewan, Canada: The role of surface casing vent flows. *Environmental Science & Technology* 57(48):19594–19601.

DOI Orphaned Wells Program Office. 2024. *Plugging Away: Documenting the Impacts of the Investing in America Agenda and Orphaned Well Clean Up Across the Country*. Esri, USGS interactive map. https://storymaps.arcgis.com/stories/5b479532e1f74356b0a84c764c5ddf34 (accessed October 25, 2024).

Dubey, M. L., A. Santos, A. B. Moyes, K. Reichl, J. E. Lee, M. K. Dubey, C. LeYhuelic, E. Variano, E. Follansbee, F. K. Chow, and S. C. Biraud. 2024. Development of a Forced Advection Sampling Technique (FAST) for Quantification of Methane Emissions from Orphaned Wells. *EGUsphere* [preprint], https://doi.org/10.5194/egusphere-2024-3040 (accessed October 24, 2024).

Environmental Defense Fund. 2021. *Plugging orphan wells across the United States*. New York: EDF Headquarters. https://www.edf.org/orphanwellmap (accessed October 25, 2024).

Environmental Protection Agency. 2024. *Waste Emissions Charge for Petroleum and Natural Gas Systems: Proposed Rule*. PowerPoint presentation given at the Technical Outreach Webinar on Proposed Regulation to Implement the Waste Emissions Charge for Petroleum and Natural Gas Systems on January 25, February 20, and March 5, 2024. https://www.epa.gov/system/files/documents/2024-01/wec-proposed-rule-technical-outreach-webinar_1.pdf (accessed November 5, 2024).

Frederick, J. 2024. *Urban Abandonment Challenges and Solutions*. PowerPoint presentation given at the Practices and Standards for Plugging Orphaned and Abandoned Hydrocarbon Wells workshop in Washington, DC, on July 19, 2024.

Gianoutsos, N.J., K.B. Haase, and J.E. Birdwell. 2024. Geologic sources and well integrity impact methane emissions from orphaned and abandoned oil and gas wells. *Science of the Total Environment* 912:169584.

Grove, C.A., and M.D. Merrill. 2022. United States documented unplugged orphaned oil and gas well dataset. U.S. Geological Survey data release. https://doi.org/10.5066/P91PJETI.

Haase, K.B., N.J. Gianoutsos, and C.C. Skinner. 2024. Measurements of water quality constituents in groundwater within 1 mile (1.61 km) of orphaned wells in the United States. U.S. Geological Survey data release. 10.5066/P91O39OU.

Haden Chomphosy, W., S. Varriano, L.H. Lefler, V. Nallur, M.R. McClung, and M.D. Moran. 2021. Ecosystem services benefits from the restoration of non-producing US oil and gas lands. *Nature Sustainability* 4:547–554.

IOGCC (Interstate Oil and Gas Compact Commission). 1964. A study of conservation of oil and gas. *Interstate Oil and Gas Compact Commission*.

Kang, M. 2024. *Session 5: Reclamation and Restoration*. PowerPoint presentation given at the Practices and Standards for Plugging Orphaned and Abandoned Hydrocarbon Wells workshop in Washington, DC, on July 19, 2024.

Kang, M., S. Christian, M.A. Celia, and R.B. Jackson. 2016. Identification and characterization of high methane-emitting abandoned oil and gas wells. *Proceedings of the National Academy of Sciences* 113(48):13636–13641. https://doi.org/10.1073/pnas.1605913113.

Kang, M., J. Boutot, R.C. McVay, K.A. Roberts, S. Jasechko, D. Perrone, T. Wen, G. Lackey, D. Raimi, D.C. Digiulio, S.B.C. Shonkoff, J.W. Carey, E.G. Elliott, D.J. Vorhees, and A.S. Peltz. 2023. Environmental risks and opportunities of orphaned oil and gas wells in the United States. *Environmental Research Letters* 18(7):074012. https://doi.org/10.1088/1748-9326/acdae7.

Kell, S. 2011. State Oil and Gas Agency Groundwater Investigations and their Role in Advancing Regulatory Reforms, A Two-State Review: Ohio and Texas. Oklahoma City, OK: Ground Water Protection Council.

Krawczyk, R. 2024. *Surface Reclamation and Restoration*. PowerPoint presentation given on behalf of Parsons Corporation at the Practices and Standards for Plugging Orphaned and Abandoned Hydrocarbon Wells workshop in Washington, DC, on July 19, 2024.

Lackey, G., H. Rajaram, J. Bolander, O.A. Sherwood, J.N. Ryan, C. Yan Shih, G.S. Bromhal, and R.M. Dilmore. 2021. Public data from three US states provide new insights into well integrity. *Proceedings of the National Academy of Sciences* 118(14).

Lackey, G., I. Pfander, J. Gardiner, O.A. Sherwood, H. Rajaram, J.N. Ryan, R.M. Dilmore, and B. Thomas. 2022. Composition and origin of surface casing fluids in a major US oil- and gas-producing region. *Environmental Science & Technology* 56(23).

Lackey, G., A. Dyer, I. Pfander, C.Y. Shih, and R.M. Dilmore. 2024. Drivers of Oil and Gas Well Integrity Issues in the Greater Wattenberg Area of Colorado [preprint]. Available at SSRN: https://ssrn.com/abstract=4955488 or http://dx.doi.org/10.2139/ssrn.4955488.

Llewellyn, G.T., F. Dorman, J.L. Westland, D. Yoxtheimer, P. Grieve, T. Sowers, E. Humston-Fulmer, and S.L. Brantley. 2015. Evaluating a groundwater supply contamination incident attributed to Marcellus Shale gas development. *Proceedings of the National Academy of Sciences* 112(20):6325–6330.

McLellan, B. 2024. *Well Plugging Prioritization - Evaluating Wellbore Integrity & Subsurface Conditions in Alaska's Orphan Wells*. PowerPoint presentation given at the Practices and Standards for Plugging Orphaned and Abandoned Hydrocarbon Wells workshop in Washington, DC, on July 18, 2024.

Merrill, M.D., C.A. Grove, N.J. Gianoutsos, and P.A. Freeman. 2023. Analysis of the United States documented unplugged orphaned oil and gas well dataset (ver. 1.1, April 2023). *U.S. Geological Survey Data Report 1167*. https://doi.org/10.3133/dr1167.

Michigan State University. *RS&GIS Aerial Imagery Archive*. https://rsgis.msu.edu/ (accessed November 5, 2024).

Pennsylvania Department of Environmental Protection, 2024. *Oil & Gas Infrastructure Investment and Jobs Act (IIJA) Project Tracker*. Esri, USGS interactive map. https://padep-1.maps.arcgis.com/apps/instant/portfolio/index.html?appid=064e373125c34182b2e132dd50d7c619 (accessed October 25, 2024).

Perrone, D., and S. Jasechko. 2017. Dry groundwater wells in the western United States. *Environmental Research Letters* 12(10):104002.

Raimi, D., J. Krupnick, J.-S. Shah, and A. Thompson. 2021. Decommissioning orphaned and abandoned oil and gas wells: New estimates and cost drivers. *Environmental Science & Technology* 55(15):10224–10230.

Rowan, E.L., M.A. Engle, C.S. Kirby, and T.F. Kraemer. 2011. *Radium content of oil- and gas-field produced waters in the Northern Appalachian Basin (USA): Summary and discussion of data*. U.S. Geological Survey Scientific Investigations Report 2011–5135.

Sandl, E., A.G. Cahill, L. Welch, and R. Beckie. 2021. Characterizing oil and gas wells with fugitive gas migration through Bayesian multilevel logistic regression. *Science of The Total Environment* 769.

Schlumberger (SLB). *Energy Glossary*. 2024. https://glossary.slb.com/en/terms/b/bridge_plug (accessed October 25, 2024).

Shaheen, S., T. Wen, A. Herman, and S.L. Brantley. 2022. Geochemical evidence of potential groundwater contamination with human health risks where hydraulic fracturing overlaps with extensive legacy hydrocarbon extraction. *Environmental Science & Technology* 56(14):10010–10019.

Soriano, M.A., H.G. Siegel, K.M. Gutchess, C.J. Clark, Y. Li, B. Xiong, D.L. Plata, N.C. Deziel, and J.E. Saiers. 2020. Evaluating domestic well vulnerability to contamination from unconventional oil and gas development sites. *Water Resources Research* 56(10).

Soriano, M.A., N.C. Deziel, and J.E. Saiers. 2022. Regional scale assessment of shallow groundwater vulnerability to contamination from unconventional hydrocarbon extraction. *Environmental Science & Technology* 56(17):12126–12136.

State of Michigan Department of Environment, Great Lakes, and Energy. *GeoWebFace Data Viewer*. https://www.michigan.gov/egle/maps-data/geowebface (accessed October 25, 2024).

Szatkowski, B., S. Whittaker, and B. Johnston. 2002. Identifying the sources of migrating gases in surface casing vents and soils using stable carbon isotopes, Golden Lake Pool, West-central Saskatchewan, In *Summary of investigations 2002 (vol. 1), 118–125*. Saskatchewan Geological Survey.

REFERENCES

Townsend-Small, A., T.W. Ferrara, D.R. Lyon, A.E. Fries, and B.K. Lamb. 2016. Emissions of coalbed and natural gas methane from abandoned oil and gas wells in the United States. *Geophysical Research Letters* 43:1789–1792.

van Oort, E. 2024. *Re-gaining Annular Isolation Using Shale/Salt as a Barrier (SAAB)*. PowerPoint presentation given at the Practices and Standards for Plugging Orphaned and Abandoned Hydrocarbon Wells workshop in Washington, DC on July 19, 2024.

Wendt, A. K., T. Sowers, S. Hynek, J. Lemon, E. Beddings, G. Zheng, Z. Li, J. Z. Williams, and S. L. Brantley. 2018. Scientist–nonscientist teams explore methane sources in streams near oil/gas development. *Journal of Contemporary Water Research and Education* 164: 80-111. https://ucowr.org/wp-content/uploads/2018/11/164_Wendt_et_al.pdf (accessed October 24, 2024).

Wilson, B. 2024. *A Modern Approach to Reclamation*. PowerPoint presentation given on behalf of Red Willow Production Company at the Practices and Standards for Plugging Orphaned and Abandoned Hydrocarbon Wells workshop in Washington, DC, on July 19, 2024.

Woda, J., T. Wen, D. Oakley, D. Yoxtheimer, T. Engelder, M.C. Castro, and S. L. Brantley. 2018. Detecting and explaining why aquifers occasionally become degraded near hydraulically fractured shale gas wells. *Proceedings of the National Academy of Sciences* 115(49):12349–12358.

Appendix A

Workshop Statement of Task

The National Academies will convene a workshop for DOI's Orphaned Wells Program Office to discuss existing practices and standards for plugging orphaned and abandoned hydrocarbon wells and will include discussion of:

- Historic and current well-plugging standards and design and operational practices used in the United States;
- How these standards and practices may differ based on factors such as well age, well depth, well location, material specification (e.g., casing, line, screening), geologic and geophysical environment, production type, distance to populated areas, and remediation and restoration requirements;
- Consideration of cost, technology, or other factors that impact the development of well-plugging plans; and
- Environmental benefits to well-plugging and/or mitigation of adverse environmental impacts from well-plugging (e.g., methane leakage mitigation, groundwater and surface water protection, surface reclamation, landscape/seascape degradation or restoration, and protection of animal and bird migration corridors).

The workshop will include perspectives from the federal government, states, tribes, industry, and other stakeholders. The workshop will include a written proceedings.

Appendix B

Workshop Agenda

WORKSHOP ON PRACTICES AND STANDARDS FOR PLUGGING ORPHANED AND ABANDONED HYDROCARBON WELLS

National Academy of Sciences Building
2101 Constitution Ave, NW
Washington, DC 20418
Room 120

JULY 18, 2024

9:00–9:15	Welcoming Remarks **Dr. Mary Feeley,** *Chief Geoscientist*, ExxonMobil (retired)
9:15–9:30	Department of the Interior: Orphaned Wells Program Office Overview **Ms. Kimbra Davis,** *Director, Orphaned Wells Program Office*, U.S. Department of the Interior

SESSION 1—ORPHANED AND ABANDONED WELL-PLUGGING: COSTS, CHALLENGES, AND BENEFITS

9:30–11:30	**Moderator: Mr. James Slutz,** *Director of Study Operations*, National Petroleum Council **Ms. Lori Wrotenbery,** *Executive Director*, Interstate Oil and Gas Compact Commission

Mr. Adam Peltz, *Director and Senior Attorney, Energy Program,* Environmental Defense Fund
Mr. David Alleman, *Director, Oil and Gas Research,* U.S. Department of Energy

11:30–12:30   Working Lunch

12:30–1:30   Compilation of State Well-Plugging and Abandoning Standards and Procedures
Moderator: Dr. Mary Feeley, *Chief Geoscientist,* ExxonMobil (retired)
Mr. Rick Simmers, *Former Chief,* Ohio Department of Natural Resources, Division of Oil and Gas Resources Management

SESSION 2—WELL-PLUGGING PRIORITIZATION: EVALUATING WELLBORE INTEGRITY & SUBSURFACE CONDITIONS

1:30–3:30   **Moderator: Prof. Mileva Radonjic,** *Professor & Samson Investment Chair in Petroleum Engineering*
Mr. Dan Arthur, *President and Chief Engineer,* ALL Consulting
Mr. Tom Kropatsch, *Oil and Gas Supervisor,* Wyoming Oil and Gas Conservation Commission
Mr. Don Hegburg, *Program Manager, Office of Oil and Gas Management,* Pennsylvania Department of Environmental Protection
Mr. Danny Sorrells, *Deputy Executive Director and Director, Oil and Gas Division,* Railroad Commission of Texas
Mr. Bryan McLellan, *Senior Petroleum Engineer,* Alaska Oil and Gas Conservation Commission
Mr. Matthew Warren, *National Oil and Gas Program Lead,* Bureau of Land Management

3:30–3:45   Break

SESSION 3—WELLBORE PROCEDURES AND BEST PRACTICES

3:45–5:15   **Moderator: Prof. Nathan Meehan,** *Professor, Harold Vance Department of Petroleum Engineering,* Texas A&M University
Mr. Drew Hunger, *President,* Seashore Petroleum, LLC
Mr. Steve Plants, *President of Abandonment Operations,* Plants and Goodwin, Inc.
Mr. James Bolander, *President,* JLB Engineering, LLC

5:15–5:30   Day 1 Summary and Preview of Day 2

JULY 19, 2024

9:00–9:05   Welcoming Remarks and Recap of Day 1
Dr. Mary Feeley, *Chief Geoscientist,* ExxonMobil (retired)

SESSION 4—ENVIRONMENTAL RISKS AND MONITORING

9:05–11:25   **Moderator: Prof. Mary Kang,** *Associate Professor, Department of Civil Engineering,* McGill University
Ms. Sarah Busch, *General Engineer,* Environmental Protection Agency
Dr. James France, *Senior International Methane Scientist,* Environmental Defense Fund
Prof. Susan L. Brantley, *Evan Pugh University Professor, Department of Geosciences Earth and Environmental Systems Institute,* The Pennsylvania State University
Dr. Greg Lackey, *Research Engineer,* National Energy Technology Laboratory
Prof. James Saiers, *Clifton R. Musser Professor of Hydrology, School of the Environment,* Yale University

11:25–12:25   Working Lunch

SESSION 5—RECLAMATION AND RESTORATION

12:25–1:55   **Moderator: Prof. Mary Kang,** *Associate Professor, Department of Civil Engineering,* McGill University
Mr. Forrest Smith, *Lead Petroleum and Environmental Engineer,* National Park Service
Mr. Ron Krawczyk, *Senior Project Engineer,* Parsons Corporation
Mr. Michael Hickey, *Program Engineer,* Colorado Energy and Carbon Management Commission
Mr. Brent Wilson, *Reclamation and Sustainability Manager,* Red Willow Production Company

SESSION 6—ADVANCES IN PLUGGING AND ABANDONMENT FOR IDLE WELLS

1:55–4:05   **Moderator: Prof. Mileva Radonjic,** *Professor & Samson Investment Chair in Petroleum Engineering*
Dr. Susan Nash, *Director of Innovation and Emerging Science,* American Association of Petroleum Geologists
Mr. Thomas Lopez, *Principal P&A,* ExxonMobil
Mr. Jesse Frederick, *Owner,* WZI

> **Mr. Nick Gianoutsos,** *Physical Scientist, Central Energy Resources Science Center,* U.S. Geological Survey
> **Prof. Eric van Oort,** *Joe J. King Chair of Engineering No. 2 and B.J. Lancaster Professorship in Petroleum Engineering,* The University of Texas at Austin

4:05–4:20   Break

4:20–4:30   Synthesis Remarks
> **Dr. Mary Feeley,** *Chief Geoscientist,* ExxonMobil (retired)
> **Workshop Staff Leads,** National Academies of Sciences, Engineering, and Medicine

4:30–4:35   Next Steps and Future Work
> **Dr. Deborah Glickson,** *Board Director,* Board on Earth Sciences and Resources

4:35   Meeting Adjourns

Appendix C

Workshop Planning Committee Member Biographies

MARY HART FEELEY (*Chair*) retired as chief geoscientist from ExxonMobil Exploration Company in 2014. Her responsibilities included advising senior ExxonMobil Upstream management on strategic geoscience matters and identifying global geoscience opportunities for ExxonMobil. Her graduate work focused on understanding depositional patterns in upper slope salt basins and the Mississippi Fan using seismic stratigraphy techniques. She also spent many years working on lease sales, prospect maturation, and energy development in the Gulf of Mexico. Feeley received a Ph.D. in oceanography from Texas A&M University. She previously served on the National Academies of Sciences, Engineering, and Medicine's Ocean Studies Board, the Committee on Guidance for NSF on National Ocean Science Research Priorities: Decadal Survey of Ocean Sciences, and the Committee on Offshore Science and Assessment for the Bureau of Ocean Energy Management.

MARY KANG is an associate professor of civil engineering at McGill University, studying methane emissions from oil and gas systems and subsurface hydrology. Kang made the first direct measurements of methane emissions from abandoned oil and gas wells in the United States, and over the past decade, she has led projects on direct measurements of abandoned/inactive wells in Pennsylvania, West Virginia, Oklahoma, California, British Columbia, Alberta, Saskatchewan, Ontario, and internationally. She conducts data mining, geospatial/statistical analysis, and machine learning to determine the scope of the emissions and develop mitigation solutions. Kang received a B.A.Sc. and an M.A.Sc. in civil and environmental engineering at the University of Waterloo, Canada, and a Ph.D. in civil and environmental engineering from Princeton University. She was a postdoctoral fellow in Earth system science at Stanford University.

DONALD NATHAN MEEHAN is a professor in the Harold Vance Department of Petroleum Engineering at Texas A&M University, specializing in carbon capture, utilization, and storage; blue hydrogen; emissions reduction in oil and gas operations; and enhanced recovery in unconventional wells using carbon dioxide. He serves as a senior technology advisor for PetroAI and as a nonexecutive director of Ignis H2, a geothermal energy startup. With more than 45 years of industry experience, he held leadership roles at CMG Petroleum Consulting; Gaffney, Cline & Associates; and Baker Hughes. Meehan served as the 2016 president of the Society of Petroleum Engineers (SPE); is a member of the National Academy of Engineering; and is a recipient of SPE's Lester C. Uren Award, the DeGolyer Distinguished Service Medal, and the SPE Public Service Award. He received the World Oil Lifetime Achievement Award and *Petroleum Economist's* Legacy Award. Meehan received a B.Sc. in physics from the Georgia Institute of Technology, an M.Sc. in petroleum engineering from the University of Oklahoma, and a Ph.D. in petroleum engineering from Stanford University.

MILEVA RADONJIC is a professor and the Samson Investment Chair in Petroleum Engineering at Oklahoma State University, where she established Hydraulic Barrier Materials and Geomimicry Labs in the School of Chemical Engineering. She spent a year at the Federation of American Scientists in Washington, DC, focusing on building materials for rapid rebuilding post-Katrina in New Orleans, prior to employment with the BP America drilling team in Houston. Her primary research interest remains focused on investigating mechanisms of rock/cement–fluid interactions and their impact on engineering performance in concrete structures, ancient monuments, and wellbores. Radonjic received a doctoral degree at the Interface Analysis Centre, University of Bristol, United Kingdom, followed by a visiting scholarship at Princeton University.

JAMES ALLEN SLUTZ is the director of study operations for the National Petroleum Council (NPC), an independent federal advisory committee to the United States, reporting to the secretary of energy. Prior to NPC, he led a global energy consulting practice with projects in North America, Asia, and Europe. Previously, Slutz served as acting assistant secretary of fossil energy at the Department of Energy (DOE) and before that as deputy assistant secretary of oil and natural gas. Prior to joining DOE, Slutz served as the Indiana oil and gas director, regulating the state's upstream oil and gas industry and natural gas storage wells. He is a former vice chair of the Interstate Oil & Gas Compact Commission. Slutz serves as an advisor to the National Bureau of Asian Research and is a board member of the local chapter of the Society of Petroleum Engineers (SPE). In his capacity with SPE, he serves as the program chair for the annual SPE/American Association of Petroleum Geologists/Society of Exploration Geophysicists Washington, DC, Technology and Sustainability Symposium. He has published papers in collaboration with the American Enterprise Institute, the East–West Center, the U.S. Chamber of Commerce Foundation, and the National Bureau of Asian Research. Slutz received a B.S. from the Ohio State University School of Natural Resources and an M.B.A. from the Ohio State University Fisher College of Business. He previously served as chair of the Committee on Earth Resources and as a member of the Board on Earth Sciences and Resources of the National Academies of Sciences, Engineering, and Medicine.